270 Current Topics in Microbiology and Immunology

Springer
Berlin
Heidelberg
New York
Barcelona
Hong Kong
London
Milan
Paris
Tokyo

Toll-Like Receptor Family Members and Their Ligands

Edited by B. Beutler and H. Wagner

With 31 Figures and 5 Tables

Springer

Professor Bruce Beutler, M.D.
Department of Immunology
The Scripps Research Institute
10550 N. Torreye Pines Road
La Jolla, CA 92037
USA
e-mail: bruce@scripps.edu, bbeutler@palm.net

Professor Dr. Hermann Wagner
Institut für Medizinische Mikrobiologie,
Immunologie und Hygiene
Technische Universität München
Trogerstr. 9
81675 München
Germany
e-mail: h.wagner@lrz.tum.de

Cover Illustration: Schematic presentation of the ten human TLRs, and their chromosomal locations in *Homo sapiens*. A molecular clock, calibrated based on the orthologous relationship between TIR domains in distantly related vertebrates (Chapter 1), has been used to date the divergence of the TLR paralogs. In all, 20 TIRs have been identified in humans. Green rectangles: leucine-rich repeats. Red rectangle, membrane-proximal leucine-rich repeat. Green oval: TIR domain. Red marks to the left of chromosomes show the approximate location of TIR-domain genes, which include the ten TLRs, members of the IL-1/IL-18 receptor group, MyD88, and MAL/Tirap.

ISSN 0070-217X
ISBN-13: 978-3-642-63975-3 e-ISBN-13: 978-3-642-59430-4
DOI: 10.1007/978-3-642-59430-4

Springer-Verlag Berlin Heidelberg New York
a member of BertelsmannSpringer Science + Business Media GmbH

http://www.springer.de

Softcover reprint of the hardcover 1st edition 2002
Library of Congress Catalog Card Number 15-12910

Cover Design: *design & production GmbH*, Heidelberg
Typesetting: Scientific Publishing Services (P) Ltd, Madras
Production Editor: Christiane Messerschmidt, Rheinau
Printed on acid-free paper SPIN: 10855994 27/3020 5 4 3 2 1 0

Preface

On occasion, the innate immune system is referred to as the "primitive" immune system. Perhaps this has dissuaded immunologists from analyzing it as energetically as they have analyzed the adaptive immune system during the past two decades. But while its phylogenetic origins are indeed ancient, and though it is "of the first type", there is nothing crude, nothing unsophisticated, and nothing "inferior" about innate immunity. On the contrary, the innate immune system has had time to achieve a level of refinement that is nothing short of dazzling, and a modicum of respect is at long last due.

Any immune system has two cardinal functions. It must destroy a broad range of pathogens, and it must spare the host. The adaptive immune system has applied a modular solution to these problems. Each cell of the adaptive immune system is prescreened to eliminate those that would produce untoward interactions with self; each cell is pre-programmed to recognize a foreign epitope that the host might one day encounter. Hence, the duties of each individual lymphocyte are quite circumscribed. And in fact, only the exceptional cell will ever be called into service. Pity the poor macrophage: it must recognize every microbe that invades the host, though some bear practically no semblance to others. On occasion, the macrophage must consume the remnants of injured host tissues, but it must never destroy healthy cells. Not content to dwell within established garrisons such as the spleen or lymph nodes, the macrophage is ever on patrol, wandering through the stroma of highly differentiated tissues and seeking its microbial quarry. The macrophage must sometimes call for assistance, enlisting the service of its myeloid relative, the granulocyte. And the granulocyte may certainly act on its own initiative.

This, at any rate, is the situation in mammals. In other organisms, innate immunity has evolved in very different directions. Every cell of a plant must provide for its own defense, for in plants, no wandering cells can come to the rescue. In insects, much depends upon proteins of the hemolymph, which perceive the presence of microbes and relay the information to diverse somatic cells, particularly within the fat body.

The cells and proteins of the innate immune system do not think, but act as though they do. They are, in effect, part of an elaborate machine that functions autonomously, but in marvelous coordination with parts of the organism that do not exhibit primary immune function. When an adaptive immune system is present, it is plain to see that it is entirely dependent upon the innate immune system that preceded it. Simply put: without innate immunity, there is no adaptive immunity.

In the last decade of the twentieth century, parallel studies in insects and mammals led to the elucidation of the principal sensing mechanism utilized by the innate immune system of each. To the surprise of many, who quite naturally held flies and humans more remarkable for their differences than for their similarities, many remnants of the defensive system shared by their common ancestor are still in evidence. More amazing still, some defensive proteins used by plants are structurally similar to those used by animals. At the heart of the evolutionary story is the TIR domain, and at the heart of microbial sensing in mammals are the Toll-like receptors of which they are a part.

In a nascent field like this one, controversy abounds, and some of the views set forth in the following chapters may conflict with others. Yet the consensus view is plain enough: the sensing mechanism of the innate immune system has quite suddenly been discovered, and new gateways to the understanding of microbial pathogenesis have been flung open. This compendium of views from some of the leaders in our attempt to understand innate immunity is intended as a milestone, marking "the end of the beginning".

March 2002

HERMANN WAGNER
BRUCE BEUTLER

List of Contents

List of Contributors

(Their addresses can be found at the beginning of their respective chapters.)

Evolution of the TIR, Tolls and TLRs: Functional Inferences from Computational Biology

B. Beutler[1] and M. Rehli[2]

The mammalian toll-like receptors (TLRs) are products of an evolutionary process that began prior to the separation of plants and animals. The most conserved protein motif within the TLRs is the TIR, which denotes Toll, the Interleukin-1 receptor, and plant disease Resistance genes. To trace the ancestry of the TLRs, it is desirable to draw upon the sequences of TIR domains from TLRs of diverse vertebrate species, including species with known dates of divergence (i.e., representatives of *Mammalia* and *Aves*) in order to establish a relationship between time and genetic divergence. It appears that a gene ancestral to modern TLRs 1 and 6 duplicated approximately 130 million years ago, only shortly before the speciation event that led to humans and mice. Though it is not represented in mice, TLR10 split from the TLR[1/6] precursor about 300 million years ago. The origins of other TLRs are more ancient, dating to the origins of vertebrate life, and some present-day vertebrate species appear to have many more TLRs than others. Moreover, the patterns of TLR expression are quite variable at the level of tissues, even among closely related species. A given TLR in species that are related by descent from a common ancestor may acquire different duties within each descendant line, so that some microbial inducers are avidly recognized in one species but not in others; likewise the intensity and the anatomic location of an innate immune response may vary considerably. In this review, we discuss the computational methods used to analyze divergence of the TIR, and the conclusions that may be safely drawn.

[1] Department of Immunology, The Scripps Research Institute, 10550 N. Torrey Pines Road, La Jolla, CA 92037, USA, *e-mail: brūce@scripps.edu*
[2] Abt. für Hämatologie und Internistische Onkologie, Klinikum der Universität Regensburg, Franz-Josef-Strauss Allee 11, 93042 Regensburg, Germany, *e-mail: michael.rehli@klinik.uni-regensburg.de*

1 Introduction

The toll-like receptors of mammals are single-spanning transmembrane proteins with ectodomains built of interspersed leucine-rich repeat (LRR) motifs. Each TLR is also characterized by a single, dominating Toll/interleukin (IL)-1 receptor (TIR) motif in the cytoplasmic domain. Viewed without prejudice, the TIR motif is a bimolecular interaction domain. Most probably, it could fulfill many functions that require bimolecular interaction, and there is no particular reason why it must be devoted to host defense. But defense seems to have been its primordial role, as both plants and animals use the TIR for defensive purposes. The TIR is conserved to such a degree that very long-range assessments of molecular evolution may be based upon it. In turn, by knowing the phylogenic context in which TIR changes occurred, it is possible to draw certain conclusions as to the function of the TLR receptors within which TIRs reside.

2 Computational Methods Applied to the Determination of Ancestry

Dwelling within the germline, DNA is commonly regarded as a molecular clock. But more accurately, it is a part of a clock. It is best likened to the clock display; mutational pressure is the mainspring that drives the clock, and natural selection is the load against which the mainspring must work. Some nucleotides may be far more subject to change than others, both because they are intrinsically more mutable, and because they are less subject to selection. But when DNA molecules of sufficient length are scrutinized, change occurs with impressive constancy. Because all organisms on earth share a common ancestor, DNA sequences residing within different species bear a true phyletic relationship to one another. Every mutational difference between two homologous sequences reflects a real event that occurred at a precise instant in time, captured and canonized by fixation within the species that sustained it. If two sequences were demonstrably derived from a common ancestral sequence, they are said to be homologues. Homologies are either orthologous (for example, TLR2 of the mouse is orthologous to TLR2 of the human), or paralogous (for example, TLR2 of the mouse is paralogous to TLR3 of the mouse).

The relationships between paralogous sequence elements that arose from recent gene duplication events are particularly thought provoking (Fig. 1). It is possible to examine large multigene families that extend across many taxa. For two paralogous genes, A and B, in two species, X and Y, we may ask: ‘Just when did A and B encode the very same protein?’ And further: ‘Was the final date of their identity before or after species X diverged from species Y?’ For surely, species X and species Y did go their separate ways at a particular

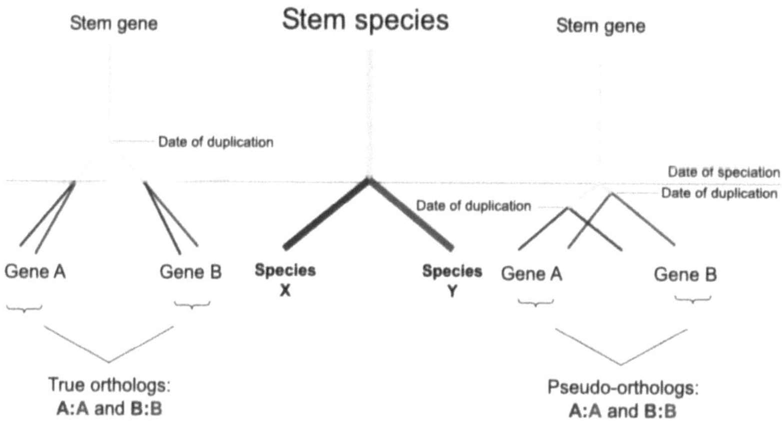

Fig. 1. The formation of paralogs, orthologs, and pseudo-orthologs. Assume the existence of two species, X and Y (*blue* and *red*, respectively), which are descendants of a single stem species (*yellow*). Further assume two paralogous genes, A and B, represented in X and Y today. Two alternative pathways may have led to this situation. In the first instance (*left*), the genes underwent duplication and divergence within the stem species, becoming de facto paralogs. When speciation later occurred, both daughter genes were maintained in the germline of both daughter species. Hence, the two modern A genes are true orthologs, as are the two modern B genes. In the second scenario (*right*), duplication events occurred independently in both of the daughter species, and the seeming orthologous relationship between members of each gene pair in each daughter species is inauthentic. 'Pseudo-orthology' can be declared if careful analysis of the four genes in question reveals closer similarity between A and B within each of the two species than between A of one species and A of the other species, or B of one species and B of the second species. As an example of pseudo-orthology, the TLR2 gene has undergone duplication in both chickens and in alligators. However, it appears to have done so long after the two species diverged from a common ancestor some 250 million years ago

moment in the past, hazy though that moment might seem to us today. And if gene duplication occurred twice after speciation – once in each descendant species – the proteins that result may closely resemble one another but are not truly orthologous. Rather, they are nacent paralogues, or at least, have the potential to become paralogues. As we shall see, such is the case with some mammalian TLRs.

The divergence of species is temporally defined by cessation of gene flow between descendants born to a common ancestral breeding population. A single individual may be the scion of two or more new species, but the bottleneck need not be so extreme, and speciation may occur even when the ancestral population is made up of thousands or millions of breeding individuals. The fossil record, interpreted in the context of geologic strata, provides the date of speciation. The divergence of genes to form paralogues is temporally defined by the date at which gene duplication occurred within an individual. This date cannot ordinarily be determined by examination of the fossil record. Rather, rates of protein evolution may be calibrated with respect to a known date of species divergence. In the example to be provided below, it is assumed that the TLR2 protein of domestic chickens and humans arose from a single primordium present in the stem species. The last common ancestor of birds and mammals is believed, within narrow limits, to have lived 310 million years ago, and to have diverged into synapsids (which gave rise to mammals) and diapsids (which gave rise to birds) (KUMAR and HEDGES 1998). Hence, the degree of change evident upon comparison of extant avian and mammalian TLR2 sequences is known to have occurred over a period of 310 million years. It may further be assumed that other TLR paralogues, having functions that are at least similar to that of TLR2, probably underwent change at similar rates. Using the TLR2 calibration and inquiring with regard to those vertebrate species that interest us, we may deduce that some paralogues of the family arose after species divergence, while others underwent duplication before species divergence.

If species X and species Y must contend with different environments and hence different selective pressures, we may begin to gain insight into the fundamental function of molecules A and B, and why paralogous evolution has been sustained – or at least we may make some educated guesses.

Some parts of the genome change with little or no constraint. For example, mutational pressure may entirely 'crush' a newly duplicated gene within 100 million years, or may render a non-functional stretch of intergenic DNA unrecognizable (LI 1997). Other parts of the genome are subject to stringent conservation, usually because they encode proteins with very specialized and indispensable functions. Such elements are the basis of molecular clocks that tick slowest. Within TLR genes, the portions that encode the cytoplasmic domains of the corresponding proteins are most stringently conserved, and yield the most reliable determinations of phylogeny occurring over long periods of time (i.e., >100 million years). Insofar as only 76% of coding changes lead to an amino acid change, amino acids provide a clock that runs slower than the nucleotides that encode them. The vertebrate TLR ectodomains change much more rapidly, and may be used for analyses of change that has occurred among more closely related species.

In Fig. 2, the amino acid sequences from the TLR2 gene of *Homo sapiens* is shown, matched against the corresponding sequences of the two extant TLR2 genes from *Gallus gallus*. Note that the two avian sequences agree with one another very closely. The genes that encode these proteins underwent duplication far more recently than the time at which birds and mammals diverged. Note also that the cytoplasmic domain of these proteins has been most resistant to change: beginning at residue 616 of the human sequence, which corresponds to the first residue of the cytoplasmic domain, and continuing through the C terminus of the molecule (residue 784) there are but 33 amino acid differences distinguishing the human and avian sequences, and no differences between the avian sequences themselves. Within the past 310 million years, only 19.5% of the residues of the cytoplasmic domain have been altered. This may be used to provide a linear standard for estimation of the times at which various present day mammalian TLRs underwent duplication, taking mathematical account of reversion and 'double hits' at individual nucleotides.

```
             1                                                  50
{galtlr2b}   ---------- ~mht*km*a* c********* ********** **********
{galtlr2a}   mfnqskqkpt mklm*qa*l* y********* ********** **********
 {humtlr2}   ---------m phtl*mv*vl gviis..*sk *essn**s** **rngi*kg*
 Consensus   ---------- ----W--W-I -TALAAHLPE EQALRQACLS CDATQSCNCS

             51                                                100
{galtlr2b}   ********** ********** ******v*** ********** **********
{galtlr2a}   ********** ********** ******l*** ********** **********
 {humtlr2}   sgs*ns**s* **ea*ks*d* sn***ty*sn s***rc***q a*v*t**g*n
 Consensus   FMGLDFIPPG LTGKITVLNL AHNRIK-IRT HDLQKAVNLR TLLLQSNQIS

             101                                               150
{galtlr2b}   ********** ********** ********** ********** **********
{galtlr2a}   ********** ********** ********** ********** **********
 {humtlr2}   t*e****s*l *s**h****y *y*sn**ss* *k**s**tf* nll**p*kt*
 Consensus   SIDEDSFGSQ GKLELLDLSN NSLAHLSPVW FGPLFSLQHL RIQGNSYSDL

             151                                               200
{galtlr2b}   ********** *********. ********** ********** **********
{galtlr2a}   ********** *********. ********** ********** **********
 {humtlr2}   **t*l**h*t k*qi*rv**m dt*tk*qrkd *a*lt**ee* e**asd*qs*
 Consensus   GESSPFSSLR NLSSLHLGN- PQFSIIRQGN FEGIVFLNTL RIDGDNLSQY

             201                                               250
{galtlr2b}   ********** ********** ********** ********d* *klafsvpek
{galtlr2a}   ********** ********** ********** ********** *eik**iene
 {humtlr2}   **k*****qn *s*l*lhmkq hillle*fv* vts*vec**l *dtd**tfhf
 Consensus   EPGSLKSIRK INHMIISIRR IDVFSAVIRD LLHSAIWLEV R---LD----

             251                                               300
{galtlr2b}   iq*lrimss* fa**islkqc l***atvpe* vsi*egmpk* m*v*mk****
{galtlr2a}   klvqnstlpl t*q*l**tga s***kyis** av**ke*rs* r***ai**v*
 {humtlr2}   se*stgetn* l***f**rnv ki**eslf** mk**nq*sg* l***fd****
 Consensus   --L------S -IKK-TF--- -FTD----QI --LL--I--L -ELE--DCTL

             301                                               350
{galtlr2b}   l*t*k*.ykq *ha.nqsq*l ri***e**s* ee****t**q svld*lslfr
{galtlr2a}   e*k*a*dmte *a*.skqs*i ***s*t*mt* ld****f**e gietqvgkl*
 {humtlr2}   n*v*nfrasd nd*vidpgkv *****rr*h* pr****y**s tlys*terv*
 Consensus   -G-G-W---- I-R-----S- ETLTI-NL-I --FYLF-DL- ----L----K

             351                                               400
{galtlr2b}   kv****t*** ****k***** l********a ***gdqs*eh ***qg*****
{galtlr2a}   *ls*as**** m***r*aryf s**l***fhd ****nnr*ge ti********
 {humtlr2}   *i******** ****l***** k********e **m*eey*kn **********
 Consensus   R-TVENSKVF LVPC-LSQHL -SLEYLDLS- NLLV---L-- SACEDAWPSL

             401                                               450
{galtlr2b}   ********** sd**m**ks* fh*r**nl** ********** **********
{galtlr2a}   ******k*** k***qaaryi sn*hk*i*** ********** **********
 {humtlr2}   ***i*r**h* a**ek**et* lt*k**t*i* **k*s*hsm* et*q***km*
 Consensus   QTLNLSQNSL -SLK-TG--L --L-NL-NLD ISENNFGEIP DMCEWPENLK

             451                                               500
{galtlr2b}   ********** ********** ********** ********** **********
{galtlr2a}   ********** ********** ********** ********** **********
 {humtlr2}   *******r*h sv*g***k** ******n*** nl*s*n**q* ****isr*k*
 Consensus   YLNLSSTQIP KLTTCIPSTL EVLDVSANNL QDFGLQLPFL KELYLTKNHL

             501                                               550
{galtlr2b}   ********** ********** ********** ********** **********
{galtlr2a}   ********** ********** ********** ********** **********
 {humtlr2}   m***d*sll* m*lvlk**** aitt****ql d**htlkt*e *gg*******
 Consensus   KTLPEATDIP NLVAMSISRN KLNSFSKEEF ESFKQMELLD ASANNFICSC

             551                                               600
{galtlr2b}   ********** ********** ********** ********** **********
{galtlr2a}   ********** ********** ********** ********** **********
 {humtlr2}   *****tqeqq al*k***d** an*l****sh ***q**qd*r **vs****ta
 Consensus   EFLSFIHHEA GIAQVLVGWP ESYICDSPLT VRGAQVGSVQ LSLMECHRSL

             601                                               650
{galtlr2b}   ********** ********** ********** ********** **********
{galtlr2a}   ********** ********** ********** ********** **********
 {humtlr2}   ***gm*cal* *l**ltg*lc hr**gl***k *m******** **rk**srn*
 Consensus   LVSLICTLVF LFILILVVVG YKYHAVWYMR MTWAWLQAKR KPKRAPTKDI

             651                                               700
{galtlr2b}   ********** ********** ********** ********** **********
{galtlr2a}   ********** ********** ********** ********** **********
 {humtlr2}   ********** r*ay****l* **e**nfn** *k******** **********
 Consensus   CYDAFVSYSE NDSNWVENIM VQQLEQACPP FRLCLHKRDF VPGKWIVDNI

             701                                               750
{galtlr2b}   ********** ********** ********** ********** **********
{galtlr2a}   ********** ********** ********** ********** **********
 {humtlr2}   ********** v*****n**k ********** ******e*** *a********
 Consensus   IDSIEKSHKT LFVLSEHFVQ SEWCKYELDF SHFRLFDENN DVAILILLEP

             751                                          795
{galtlr2b}   ********** ********** ********** ********** *****
{galtlr2a}   ********** ********** ********** ********** *****
 {humtlr2}   *ekk***q** ********** ******m**a *reg**v**r **i**
 Consensus   IQSQAIPKRF CKLRKIMNTK TYLEWPPDEE QQQMFWENLK AALKS
```

Fig. 2. Amino-acid alignment of two chicken TLR2 homologs (galtlr2a and galtlr2b) with human TLR2 (humtlr2). Alignment was accomplished using the GCG programs Pileup (default settings) and pretty (with consensus shown in the *bottom line*, *in caps*, and differences from the consensus shown by *asterisks*)

3 Ascertainment of a Profile: Bootstrapping to Identify all TIRs in Nature

The majority of the cytoplasmic domain of all of the TLRs is part of the evolutionarily conserved TIR domain. A straightforward search for homology among translated nucleotide sequences, performed using the Tblastn algorithm which converts all nucleotide sequences in a database into amino acid sequences in all six reading frames and searches for homology to an amino acid query sequence, is sufficient to identify most of the TIRs that have been captured by DNA sequencing across all taxa.

Using the human TLR2 cytoplasmic domain sequence as a query, a large number of other, homologous sequences are immediately recognizable, and are spread across many taxa. Without drawing an immediate conclusion as to whether the most distant of homologies spotted are authentic, a more reliable query sequence may be produced by trimming the cytoplasmic domains of the most 'trustworthy' homologues, and then aligning these domain sequences using the GCG program Pileup.

To realize a broad and fairly-weighted assortment of TIRs with which to generate an alignment and then a profile, a species for which complete genomic sequence is known should ideally be used. *Homo sapiens* and *Drosophila melanogaster* fulfill this requirement, and all TIR-bearing sequences from these two species can be used without fear of duplication (which might cause unwanted 'weighting' if TIRs from several mammalian species were included) or omission. The TIR sequences from human TLRs 1-10, the IL-1 and IL-18 receptor and receptor-like chains, the orphan receptors ST2 and SIGIRR, and the transducers MyD88 and MAL (also known as TIRAP) (Fitzgerald et al. 2001; Horng et al. 2001) are obtained from the genomic sequence of *Homo sapiens*. The TIR sequences of Tolls 1–9 and MyD88 are obtained from the genomic sequence of *Drosophila melanogaster*.

A Fasta alignment is performed on all of these proteins, using the cytoplasmic domain of *H. sapiens* TLR2 as a query. Using default stringency settings, not all of the TIRs are recognized by this procedure, for some have diverged from TLR2 beyond recognition. However, the alignments that do result are a 'stepping stone' to the past. They include: human TLRs 1, 2, 4, 5, 6, 7, 8, and 10; human MyD88 and MAL; and human IL-18R1, *Drosophila* Tolls 1, 5, 8, and 9. All 15 recognized sequences can be used to create an optimal sequence *alignment* using the program Pileup (Fig. 3), with maxgap = 3,000, and with a gap creation penalty of six (i.e., somewhat relaxed stringency).

From this alignment, an idealized sequence *profile* can be deduced using the program Hmmerbuild, which creates a hidden Markov model of the alignment generated by Pileup. The profile is an imaginary sequence, which does not exist in nature. It represents the idealized consensus of all sequences fed into the alignment. It may be used with the program Hmmersearch to find all protein sequences in a database such as PIR, Genembl, or Swissprot that match the profile.

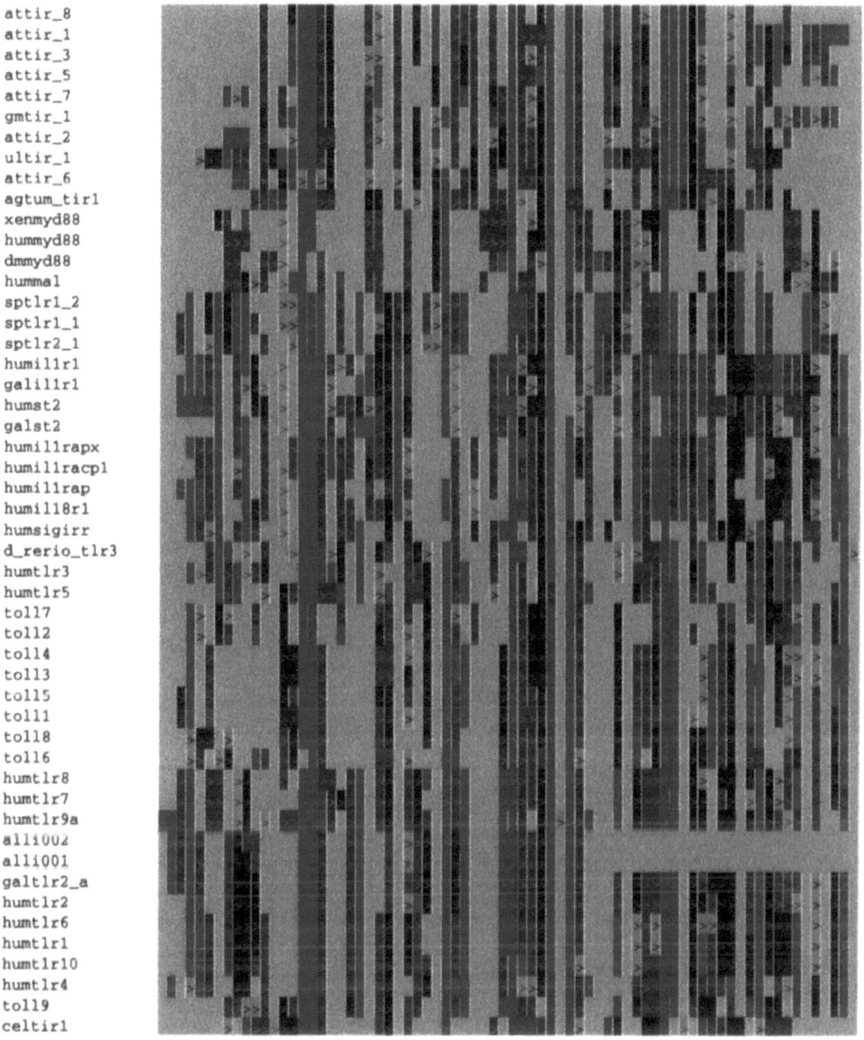

Fig. 3. Alignment of TIR domains across more than 1 billion years of evolution reveals cardinal features that define the motif. In this scheme, residues are not legible. However, hydrophobic residues are shown in *magenta*, basic residues in *dark blue*, acidic residues in *green*, aromatic residues in *red*, glutamine and histidine in *cyan*, glycine in *orange*, and proline in *black*. The 'toll' designations refer to *Drosophila* proteins, and 'tlr' designations refer to other species. st2, ilr, il18r, and sigirr designations are self-explanatory. *at*, *Arabidopsis thaliana*; *ul*, *Linum usitatissimum* (flax); *gm*, *Glycine max* (soybean); *agtum*, *Agrobacterium tumefaciens*; *xen*, *Xenopus laevis*, *dm*, *Drosophila melanogaster*; *hum*, *Homo sapiens*; *gal*, *Gallus gallus*; *alli*, *Alligator mississippiensis*; *cel*, *Caenorhabditis elegans*; *sp*, *Strongylocentrotus purpuratus*; *d_rerio*, *Danio rerio*

The alignment and profile generated in this manner utilize the most essential features of 15 TIR-bearing proteins from two species separated by at least 600 million years of evolution. The profile, used as a query sequence, illuminates the dark recesses of the evolutionary past, and permits the acquisition of TIRs that are more distant still. The most reliable of these TIRs, in turn, may be used to make a fresh alignment and profile, crossing the evolutionary divide between animals and plants, and if applied to a database that is all-encompassing, elucidating every TIR domain that has been acquired.

3.1 Treemaking: the Descent of the TIR

Using the calibration standard provided by the divergence between the TIR of TLR2 from *Homo sapiens* and *Gallus gallus*, an unrooted tree of TIR divergence can be constructed by means of maximum parsimony (the program Paupsearch, with a phylogram drawn by Paupdisplay, set for mean distance; Fig. 4), to reveal clusters of evolutionarily similar sequences that bespeak the explosion of diversified TIR structures that arose whenever a new cytoplasmic domain was found 'useful' according to the criteria applied by natural selection. Parsimony is a method in which computation is designed to identify the tree in which the least amount of evolutionary change explains the data that are observed (i.e., the minimum of total branch length to explain all extant protein species).

The TIR domains represented in this tree are derived from five distant vertebrate classes (*Osteichthes*, *Amphibia*, *Reptilia*, *Aves*, and *Mammalia*), the phylum *Echinodermata*, the insect order *Diptera*, the kingdom *Viridiplanta*, and the superkingdom *Eubacterium*. The unrooted tree does not address speciation. Rather, it shows the divergence of protein structure as it occurred irrespective of the germline that harbored it.

Note first that a tripartite division of the tree is apparent, with separate branches leading to animals, plants, and bacterial TIR representatives, and though the tree is unrooted, it would be reasonable to assign this tripartite branch point as the actual 'root.' In interpreting divergence dates, it is necessary to make use of a time standard based upon a single class of protein, and to assume that similar evolutionary pressures drove divergence of this class of proteins in all life forms, at all times. Hence, the TIRs from TLR proteins are used for this purpose.

A time standard may be based upon the fossil record (Fig. 5A), which dates the divergence of lobe-finned fish and the progenitors of land-dwelling vertebrates at approximately 370 million years ago, the divergence of the progenitors of birds (synapsids) from the progenitors of mammals (diapsids) at approximately 310 million years ago, and the divergence of the progenitor of modern birds and crocodilians (*Euparkeriidae*) at approximately 250 million years ago. The branch lengths for divergence are in rough proportion to the fossil record, and a linear estimate of paralogue gene duplication dates may be made accordingly (Fig. 5B).

The split that produced TLR1 and TLR6, for example, with its mean branch length of 9.5, corresponds to a divergence date of 130 million years ago. This event occurred immediately prior to the speciation event that led to modern humans and mice, insofar as the TLR1 and TLR6 are true orthologues in humans and mice, but only just so. Most certainly the divergence of TLR1 and TLR6 occurred after the split between eutherian and metatherian mammals (reliably dated at 170 million

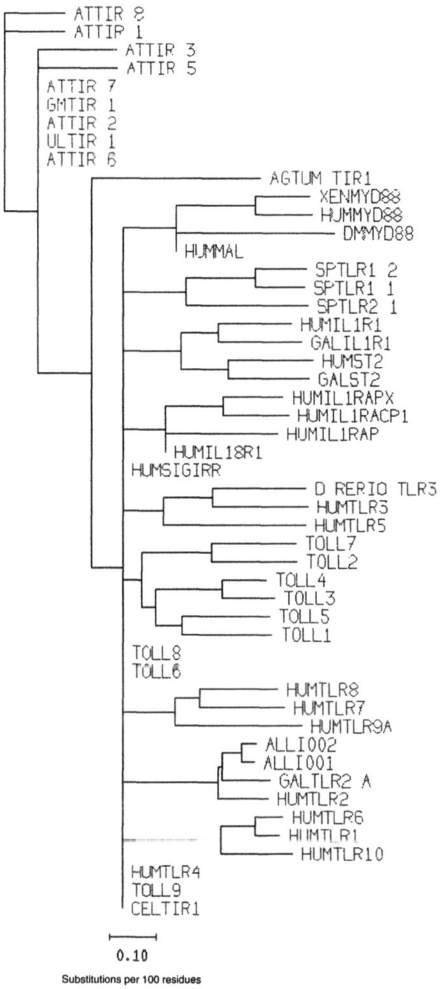

Fig. 4. Evolutionary relationship between TIR domains across the phylogenetic tree. Abbreviations are as in Fig. 3. Note that tree is unrooted, and branch lengths (in the horizontal dimension) are drawn in close proportion to the elapse of time since divergence

years ago). The divergence of the TLR1/6 precursor from what became human TLR10 probably occurred about 300 million years ago. Interestingly, TLR10 has been lost from the mouse genome, which exhibits only TLRs 1 and 6: a pair of TLR genes in tight linkage with one another, where humans show a triad of TLR genes.

Some of the Toll divergences (e.g., the split between Toll-3 and Toll-4) that occurred within the ancestors of modern *Drosophila* may have taken place contemporaneously with the rise of vertebrates. The TLR7/8/9 cluster arose earlier still: close to 500 million years ago. But other divergences were truly ancient, and probably occurred well before the dawn of vertebrate life. Mammalian TLR4 and *Drosophila* Toll-6, Toll-8, and Toll-9 appear to have arisen from such ancient events.

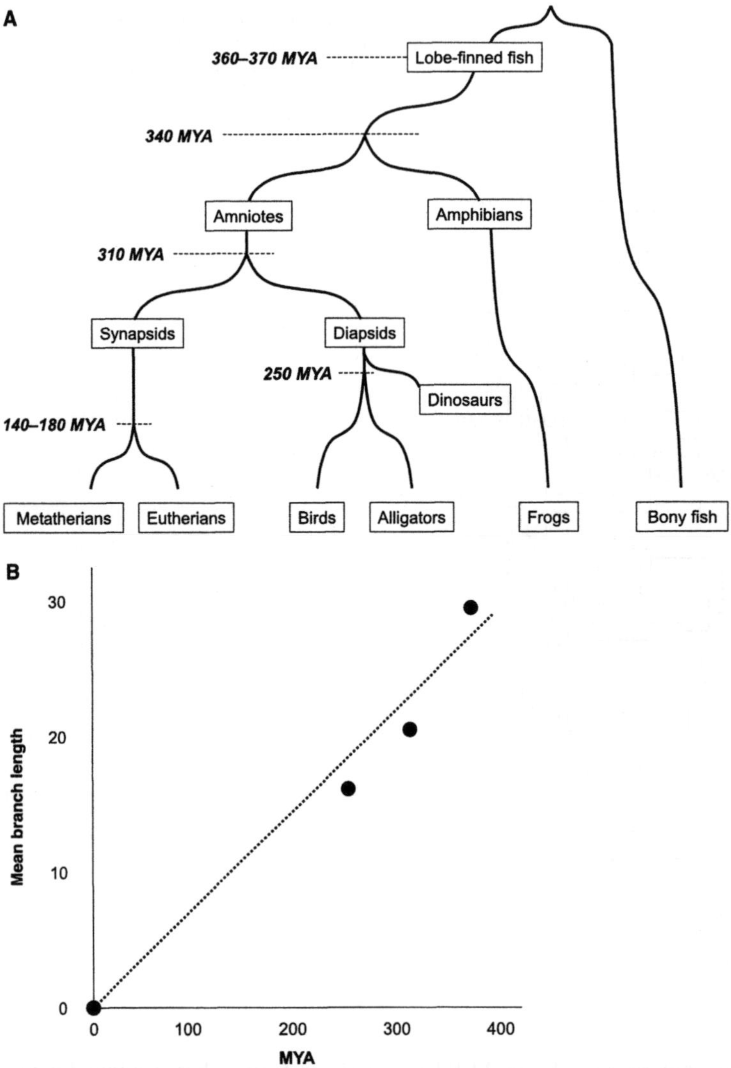

Fig. 5A,B. Evolution of vertebrate life, and the TIR domain of vertebrate TLRs. **A** The descent of metazoans is traceable in the fossil record, and certain divergence points (*bold*) can be dated with precision based on geologic data. **B** These dates may be used to apply an absolute standard for ascertainment of the time at which the divergence of specific TIR sequences began, based on the branch length in divergence trees like the one shown in Fig. 3. The only data points available, at present, include the divergence of fish TLR3 from mammalian TLR3 (approximately 360 million years ago, *MYA*); the divergence of bird TLR2 from human mammalian TLR2 (approximately 310 million years ago); and the divergence of bird TLR2 from alligator TLR2 (approximately 250 million years ago). The dates of divergence of TIR domains from one another may be interpolated or extrapolated from the line

Other TIRs, residing within different proteins, may have evolved at different rates. Hence, the TIRs that comprise the cytoplasmic domains of the IL-1R/IL-18R/ST2/SIGIRR cluster must be regarded separately. Similarly, MyD88 and

MAL must be considered on their own. Insofar as amniotes diverged from amphibians approximately 340 million years ago, and the branch length of the human: *Drosophila* MyD88 divergence is 61.5 while the branch length of the human: *Xenopus* MyD88 divergence is 18, it might be surmised that the common ancestor of humans and fruit flies lived approximately 1.2 billion years ago. However, a more recent divergence is generally accepted, based on the analysis of a larger number of proteins, some of them more highly conserved and capable of furnishing 'slower clocks.'

Where plant and bacterial TIRs are concerned, a true ancestral relationship may be contemplated, as the bacterial TIR is an outgroup, distanced from both the plant and animal clusters. However, horizontal exchange of genetic material cannot be excluded, as the rate of TIR evolution in bacteria might be very different from that in eukaryotic environments. In addition to *Agrobacterium tumefaciens*, other bacterial species also exhibit TIR-bearing proteins, not shown in the tree for the sake of simplicity. The reason for their retention in the bacterial genome is unclear. One possibility, however, may be found in the fact that the type III secretion system of certain bacteria permits the transfer of proteins from the bacterial cytosol to the cytoplasm of a mammalian host cell. Among the proteins transferred in this manner, the TIR bearing proteins may serve a 'decoy' function, confounding the immune response of the host. Much the same may be said of TIR-like sequences that are found in *Vaccinia virus* (Bowic et al. 2000), and in this case, it can hardly be doubted that horizontal transfer occurred, insofar as viruses are believed to originate within hosts.

It is important to remember that there may be many other branches on this unrooted tree: branches yet undiscovered. Note that the three known TIRs of the purple sea urchin (*Strongylocentrotus purpuratus*) are rather similar to one another, but that all diverged from the mammalian and dipteran TIRs long ago – perhaps at a moment in time prior to the divergence of mammals and flies. But because echinoderms and mammals are believed to have diverged more recently than flies and mammals, the common ancestor of flies, mammals, and sea urchins must have borne still other types of TIR, ancestral to the modern sea urchin TIRs. Even if all extant species were to be examined, some ancestral TIR genes might never be found. They may have ceased to exist in the modern world, crushed by mutational pressure as specific selective forces were relaxed. Will all of the human TLRs be represented in our descendants many millions of years hence? It is, of course, impossible to say. But there is no reason to suspect that only three or four TIR types were present in our remote ancestors: rather, they may have had many more.

3.2 Separate Considerations of the TIR and the TLR Ectodomains

The TIR, as mentioned at the beginning of this chapter, is a bimolecular interaction motif. It is found in the TLRs, IL-1-related receptors, and transducer proteins that serve them. But a receptor-oriented function may not always have been the rule, and in modern plants, the TIR does not reside within receptors at all (Chap. 2, this volume).

In general, it may be suggested that receptors evolve 'from the inside out,' often beginning as enzymes or structural proteins within the cytosol, and ending as full-fledged signal transducers (Fig. 6). A cytoplasmic enzyme may become anchored to the membrane because this distribution affords a selective advantage. Regulation of enzymatic activity is commonly advantageous, and this may apply to the membrane-bound enzyme just as to its cytosolic precursor. Through exon shuffling or other types of mutation, an ectodomain with preference for a given ligand develops. Ultimately, a cytoplasmic enzyme activity becomes subject to the concentration of specific molecule(s) in the extracellular milieu.

The modular design of receptors, and the adoption of a handful of transduction mechanisms by a huge number of plasma membrane receptors, speaks in support of this hypothesis. Moreover, from a practical standpoint, it offers the opportunity for sudden and drastic changes in ligand specificity. As has previously been suggested, co-evolution of a family of ligands and receptors may proceed through duplication of the receptor gene, the acquisition of a novel function by one of the daughter genes, and finally, duplication of the ligand gene (Bazzoni and Beutler 1996). Moreover, certain ligands [e.g., nerve growth factor (NGF)], have been 'captured' by receptors in the course of evolution: though not a member of the tumor necrosis factor (TNF) family of ligands, NGF was appropriated by a member of the TNF receptor family at some point during vertebrate evolution.

The Tolls and TLRs provide an excellent opportunity for the retrograde analysis of evolutionary events, insofar as all forms of capture and modification of function seem to have toyed with the system since the early days of metazoan evolution. We may begin with the assumption that the primordial function of the TIR was related to immune defense. This seems to be the case in most taxa today, and the appropriation of Toll (Toll-1) for use in *Drosophila* development seems to have been a departure from the status quo ante.

Four components of the same system of signaling proteins (Spätzle, Pelle, Tube and Cactus) are utilized by both the developmental and defensive pathways. Hence, it was not an adaptation of Toll per se that was required to co-opt the

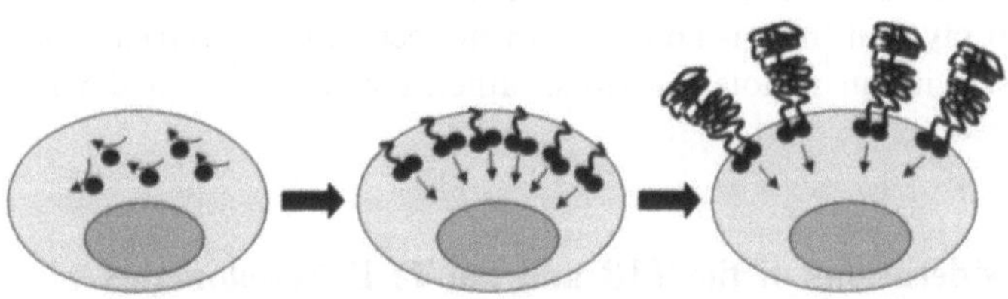

Fig. 6. Schematic representation of the evolution of a plasma membrane receptor from a cytoplasmic precursor. As stated in the text, an enzyme may achieve membrane localization, and subsequently develop an ectodomain that confers responsiveness to the external milieu

pathway for development. Rather, the switch to development must have seized upon changes at both the proximal and distal ends of the core pathway. One change must clearly have involved the development of a novel proteolytic pathway permitting cleavage of Spätzle. This might, perhaps, have involved only a single mutation, insofar as the developmental cascade (involving nudel, windbeutel, gastrulation defective, snake, and easter) might have been pre-formed, and may, indeed, have existed to fulfill other developmental functions. Moreover, the development of dorsal, a homologue of dif, may be viewed as a relatively straightforward event. More problematic is the question as to how multiple genes that confer a ventralizing effect on embryo polarity came under the control of the novel transcription factor, dorsal. It must be supposed that a wide variety of physical forms bridged the gap between present-day flies and the stem species to which *Drosophila* and vertebrates are related. Indeed, this is known to have been the case.

The overall sequence of events, then, may have been as follows:

1. The original Toll bound microbial ligands.
2. Spätzle began to respond to microbial ligands such as fungal products and peptidoglycan, which served to cleave the protein through a pathway yet unknown, for original purposes yet unknown.
3. Spätzle became an activating ligand for Toll.
4. A developmentally active proteolytic cascade 'co-opted' Spätzle.
5. A novel Rel-family transcription factor (dorsal) arose and diverged to acquire developmental specificity.
6. Toll relinquished contact with microbial ligands altogether.

3.3 The Ectodomain Tree: a Faster Clock, Driven by Ligand Preference

In keeping with the premise that receptors, once established, can undergo radical changes in ligand specificity through domain swapping, the use of an ectodomain based on immunoglobulin-type repeat motifs was adopted to form the IL-1, IL-18, ST2, and SIGIRR receptors. While the function of the last two members of the group is unknown, the IL-1 and IL-18 receptors have been studied in great detail. It is known that these receptors reside on cellular components of the adaptive immune system, including T lymphocytes. The cytokines that occupy these receptors serve primarily as a 'bridge' between the innate immune system, which produces them, and the adaptive immune system, which receives their signals. IL-1, for example, is an important cofactor for T-cell activation, triggering the production of IL-2, which in turn exerts a mitogenic effect. The costimulatory effect of IL-1 in IL-2 induction was the assay upon which purification of IL-1 was based (Gery et al. 1972). IL-18, by contrast, assists in the induction of interferon-γ, and was formerly known as interferon gamma inducing factor. It abets the Th1 response of T cells, and by way of a reflex arc involving interferon-γ, activates the macrophages that produce it.

Although invertebrate organisms have certainly had the opportunity to devise receptors in which the TIR is coupled to receptors with other types of ectodomain structure, no examples are known as such. Rather, in most organisms, a leucine-rich repeat motif has been preferred, as in the vertebrate TLRs. But as pointed out elsewhere, the structure of the ectodomain has evolved more quickly than the structure of the TIR. This presumably reflects the demands of a changing environment, in which different forms of selective pressure have come to dominate the field over time. In the case of the extant mammalian TLRs, the selective pressure in question was almost certainly microbial. However, of the nine *Drosophila* Toll receptors, only one (Toll, or Toll-1) is capable of mediating any response to infection, and at that, it does not represent a direct molecular interface with any component of the microbes to which it responds.

What, though, of the TLR ectodomain in vertebrates? Why has the TLR repertoire been all but abandoned in fish [only a single representative can be found in the *Danio rerio* expressed sequence tag (EST) database], though preserved and in some cases clearly expanded in mammals? There are few protein families that exhibit quite this level of plasticity, and in all likelihood, environmental differences are responsible. Where individual orthologues are represented in multiple species, the ectodomain clock seems to run very rapidly indeed.

The TLR with the best-characterized ligand is TLR4. TLR4 is highly specific in its ability to recognize lipopolysaccharide (LPS). TLRs 2 and 4 diverged from a common primordium approximately 500–600 million years ago: at a time rather near to the beginning of vertebrate evolution, and near to the time of divergence of Toll-9 from the precursor of vertebrate TLRs 1, 2, 4, 6, and 10. But remarkably, though TLR2 is represented in modern birds, reptiles, and amphibians, TLR4 has so far escaped detection in the genome of any non-mammalian vertebrates. The genomes of the zebrafish (*Danio rerio*) and two puffer fish genera (*Tetraodon nigroviridis* and *Takifugu rubripes*) are nearing completion, and none yet displays orthologues of the mammalian TLRs (only a *Danio rerio* EST sequence with homology is known). In the amphibian *Xenopus laevis*, only a MyD88 orthologue is represented to date, and no receptors have been discovered. In the domestic chicken, *Gallus gallus*, only a pair of TLR2 derivatives have been identified so far, very recently diverged from one another (Fukui et al. 2001), though its genome is far from complete. Unpublished studies (I. Smirnova, J. Mann, A. Poltorak and B. Beutler) have revealed a pair of TLR2 orthologues in the American alligator (*Alligator mississippiensis*): clearly pseudo-orthologues with respect to the *Gallus gallus* pair.

Extreme LPS sensitivity is quintessentially mammalian – at least in adult organisms (Libby et al. 1986), and one is tempted to conclude that LPS sensing via TLR4 was an experiment that was carried out only in mammals. The duplication product that ultimately became TLR4 was, at one stage or another, probably lost in other vertebrate lines or their progenitors. Among extant mammalian genera, great variability in LPS sensitivity is evident. Hence, even now, TLR4 might prove quite dispensable in some mammals, though surely not in all of them.

It is worth noting that other organisms have found TLR-independent solutions to the problem of LPS detection. Best known, perhaps, is the horseshoe crab,

Limulus polyphemus, in which LPS-sensitive amoebocytes support a proteolytic cascade that detects LPS with great sensitivity. While a TLR is indeed expressed by a relative of *Limulus*, the Japanese horseshoe crab *Tachypleus tridentatus*, the protein is not known to be essential for LPS responses (Inamori et al. 2000).

4 Promoter Divergence and Variations of TLR Expression Among Species

To this point, we have only considered those parts of a TLR gene that code for its proteins. As discussed above, a high percentage of nucleotide exchanges translate directly into amino acid changes that possibly affect ligand specificity (in the case of the ectodomain) or down-stream signaling (TIR domain). Away from the coding regions of a given gene locus, evolutionary changes generally occur more rapidly over time, with the exception of regions of a locus that control expression.

We may assume that the regulation of a given gene at the level of transcription is an important aspect of its function – in particular its cell-type specificity and its ability to be induced or down-regulated by exogenous or endogenous stimuli. Provided that the transcriptional machinery was conserved, which is likely to be the case for the two mammals *Homo sapiens* and *Mus musculus*, one may also wish to analyze changes in regulatory sequences that occurred during evolution. This, however, is far more complicated than the analysis of coding transcripts. In most cases, regulatory sequence regions must be defined experimentally – the position and function of sequences that control the expression of a gene cannot yet be deduced in silico. With the exception of proximal promoters, the position of regulatory elements is not consistently circumscribed. Relatively small regulatory sequences can be distributed over a large locus region and currently it is not possible to predict these regions reliably by computational means. Even the most readily available regulatory sequence, the proximal promoter, including the core promoter and adjacent enhancer elements, has to be defined by an exactly determined transcription start site. To date, only limited information on transcription start sites of human or mouse TLRs is available, and only two TLR promoter regions, those of TLR2 and TLR4 genes, have been defined and analyzed in both mouse and man (Musikacharoen et al. 2001; Rehli et al. 2000; Wang et al. 2000, 2001).

4.1 Expression and Regulation of Human and Murine TLR4 Genes

The sequence and structure of TLR4 genes in humans and mice have been determined experimentally (see Fig. 7C,F). The predominant TLR4 expressing cells in both species are of myeloid origin (monocytes, macrophages, myeloid dendritic cells and granulocytes) and the proximal promoter of TLR4 genes, which is conserved between man and mouse, contains features typical of a myeloid-specific gene (REHLI et al. 2000). These include the absence of a TATA-box and the presence of several purine-rich sequences that are recognized by the myeloid- and B-cell specific transcription factor PU.1.

In human tissues, TLR4 is expressed most strongly in spleen and peripheral blood lymphocytes (PBL), weaker expression levels were detected in most other tissues with the exception of liver, where TLR4 mRNA was not detected (ROCK et al. 1998). In mice, however, TLR4 is most strongly expressed in lung, heart, and spleen, followed by muscle, liver, and kidney, indicating that the basal tissue expression is not identical, despite the presence of homologous promoter regions (QURESHI et al. 1999a,b). Interestingly, expression levels of TLR4 transcripts are modified differently upon stimulation with the major TLR4 agonist LPS: in human monocytes/macrophages, TLR4 expression increases after activation (MUZIO et al. 2000), whereas in murine macrophages, TLR4 transcripts are down-regulated after cellular activation (POLTORAK et al. 1998). A comparison of TLR expression is shown in Table 1. The mechanisms responsible for these differences still need to be identified; however, they are probably due to differences in regulatory sequences in mouse and man.

4.2 Expression and Regulation of Human and Murine TLR2 Genes

In contrast to human and murine TLR4 genes, TLR2 promoter sequences in both species diverged during evolution. Using the ClustalW algorithm (or similar programs to detect sequence homologies) no significant level of homology between human and murine 5′-upstream sequences or 5′-untranslated regions (UTRs) can be detected (Fig. 7D). A non-coding copy of exon III is located approximately 4 kb upstream of the human TLR2 gene. The presence of this pseudogene indicates that in man, a duplicate of the coding exon III may have been placed into a different regulatory context during evolution.

The different regulatory sequences provide the most likely explanation for the observed differences in TLR2 tissue distribution, e.g., expression of TLR2 in murine, but not in human T cells or thymus. In mice, TLR2 expression is low or

Fig. 7A–F. Human and murine TLR gene and promoter structures. Putative transcription factor binding sites in proximal promoter regions of TLR genes (**A–C**) are indicated as predicted by the MatInspector Professional tool. Translated exon sequences are shown as *gray* boxes. Transcription start sites were determined by RNA-ligase mediated PCR or primer extension (MUSIKACHAROEN et al. 2001; REHLI et al. 2000; WANG et al. 2001) (see also V. Haehnel, unpublished results; also M. Rehli, unpublished results) and are indicated by *arrows*. A comparison of gene structures of TLR2, 3 and 4 is also shown (**D–F**). Proximal promoter regions (*light gray*) and exons (non-coding sequences, dark *gray*; coding sequences, *colored*) are shown as *boxes*. Note that both TLR3 genes (**E**) contain a unique intron that divides the TIR domains coding sequences. Non-coding 5′-exons as well as promoter regions are different in the two TLR2 (**D**) or TLR3 (**E**) orthologs. Proximal promoter regions of TLR4 orthologs (**F**) are homologous. However, both genes acquired different additional exons (human exon II, mouse exon III), which are subject to alternative splicing (IWAMI et al. 2000; REHLI et al. 2000)

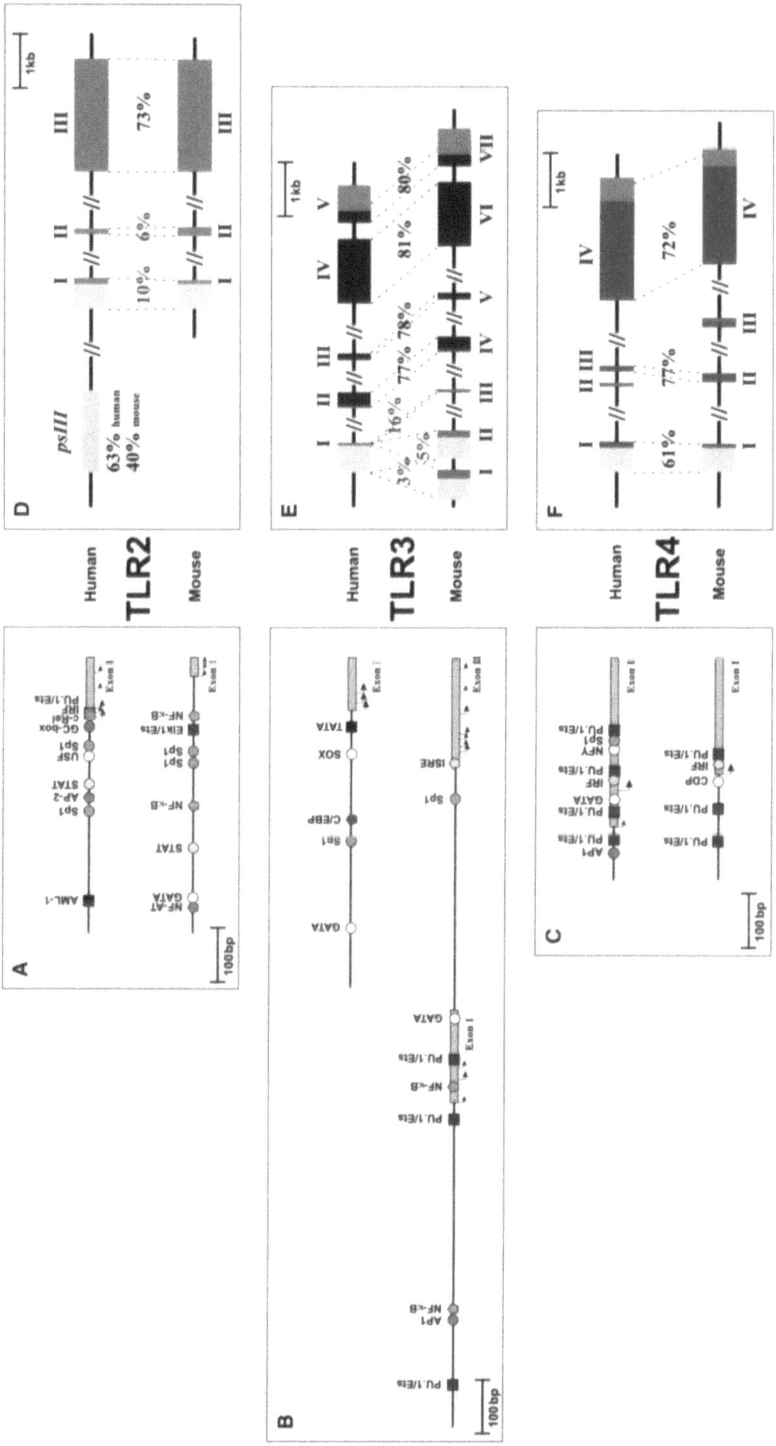
A
B
C
D
E
F
Human
TLR2
Mouse
Human
TLR3
Mouse
Human
TLR4
Mouse
100bp
1kb
psIII
63% human
40% mouse
73%
81%
80%
72%
77%
61%

Table 1. Expression of TLR2, 3 and 4 in human and mouse tissues

		Brain	Heart	Kidney	Liver	Lung	Spleen	Thymus	PBL	T cells	B cells	NK cells	MO/MAK	Myeloid DC	MO/MAK + LPS
TLR2	Human						+	–	+ + +	–	–	–	+ +	+	→/↗
	Mouse	+	+/–	+/–	+/–	+ + +	+ + +	+ +	+/–	+			+		↑↑↑
TLR3	Human	–	+/–	+/–	+/–	+/–				–	–	–	–	+ +	→/↘
	Mouse	+/–	+/–	+/–	–	+ +	+/–	+/–					+		↑↑↑
TLR4	Human	+/–	+/–	+/–	–	+/–	+ + +	+/–	+ + +	–	+/–	–	+ +	+	↑↑
	Mouse	–	+ + +	+/–	+/–	+ + +	+ +			+/–			+ +		↓↓

Expression data are based on Northern blot and PCR data from: Alexopoulou et al. 2001; Chaudhary et al. 1998; Matsuguchi et al. 2000a,b; Medzhitov et al. 1997; Qureshi et al. 1999a,b; Rock et al. 1998; Visintin et al. 2001; Yang et al. 1998). Levels of mRNA expression are indicated as follows: –, not detected; +/–, low; +, + +, or + + +, moderate to strong. Arrows indicate an increased or decreased expression following LPS treatment of monocytes/macrophages (*MO*/*MAK*).

undetectable in blood cells and is strongly induced by pro-inflammatory cytokines or microbial activators (e.g., LPS) (MATSUGUCHI et al. 2000a,b). Accordingly, binding sites for nuclear factor (NF)-κB and STAT5 in the murine TLR2 promoter (see Fig. 7A) have been implicated in the rapid induction of TLR2 expression by pro-inflammatory stimuli in murine macrophages. In humans, however, highest constitutive levels of TLR2 expression have been observed in PBL. In human monocytes, TLR2 expression is further up-regulated after adherence to tissue culture plates, but no additional induction was observed after short-term stimulation with bacterial products. In contrast to the murine TLR2 promoter, the human promoter is not induced by microbial agonists (e.g., LPS) in monocytic cell lines.

4.3 Expression and Regulation of other TLR Genes

The available information on expression and regulation of other TLRs is limited. Neither transcription start sites nor proximal promoter regions have been defined for TLR genes other than TLR2 or TLR4 and published expression and sequence data are mostly insufficient to allow a comparative analysis of gene expression in mouse and man. In some cases, however, complete 5'-UTR of TLR transcripts may be available from Genbank for sequence comparison between species. By comparing 5'-UTR sequences as well as immediately upstream genomic sequences, which represent putative promoter regions, one is able to determine their degree of conservation and possibly explain similarities or differences in expression patterns.

This can be done for human and murine TLR3 transcripts. Following the determination of transcriptional start sites for TLR3 orthologues which actually map the published 5′-UTR sequences (M. Rehli, unpublished observation), human and mouse genome sequences can be searched for upstream genomic regions using the Blastn algorithm. A subsequent screen for putative transcription factor binding sites can be performed using the MatInspector Professional program (Genomatix Software) based on previously defined consensus sequences. The resulting structure of putative promoter region for both TLR3 genes is shown in Fig. 7B. As described for TLR2 orthologues, human and murine TLR3 genes appear to utilize different regulatory regions to control their expression. In humans, TLR3 expression is restricted to myeloid dendritic cells (MUZIO et al. 2000); expression in other major blood cells (including monocytes, granulocytes, natural killer cells, T cells and B cells) and macrophages is undetectable at the transcript level. In contrast, TLR3 is expressed and strongly induced in murine macrophages (and tissues) after LPS treatment, which is not the case in humans (ALEXOPOULOU et al. 2001; MUZIO et al. 2000). Non-coding 5′-exons and putative upstream promoter regions are completely different in both species. The presence of consensus binding sites for NF-κB within the murine, but not the human promoter region may explain the different inducibility of TLR3 in the two species.

As mentioned above, the available sequence and expression data for other TLRs is not sufficient to determine putative regulatory regions. However, certain lines of evidence suggest that other TLR genes may also be regulated differentially in mouse and man. Expression of TLR9 transcripts, for example, may be regulated in a species-specific manner: in humans TLR9 is primarily detected in plasmacytoid dendritic cells (Kadowaki et al. 2001; Krug et al. 2001), whereas in mice, macrophages and myeloid dendritic cells express this receptor (Hemmi et al. 2000).

The observed differences in the regulation of murine and human TLR genes suggest that in regulatory regions a number of changes occurred in a short time-frame (since the divergence of mice and humans). For some genes (e.g., TLR4), those changes may have been more subtle, others may have been placed into a completely different regulatory context (e.g., TLR2 and TLR3). This surely widened the spectrum of possible modifications that were demanded by the selective pressure of a rapidly changing microbial environment. Evolution not only changed the structure of extracellular TLR domains, but also modified cellular expression patterns, in particular the cell type-specificity of a TLR and a cell's ability to induce expression upon activation – a feature that may be important for this family of pattern recognition receptors. A more detailed knowledge of regulatory sequences of mammalian TLR genes may help to explain many species-specific responses to microbial patterns.

Acknowledgements. The authors are grateful to A. Poltorak and I. Smirnova for contributing partial sequences of the TLR2 genes from *A. mississippiensis* for analysis. This work was funded by a grant from the NIH (GM60031-03).

References

Alexopoulou L, Holt AC, Medzhitov R, Flavell RA (2001) Recognition of double-stranded RNA and activation of NF-κB by Toll-like receptor 3. Nature 413:732–738

Bazzoni F, Beutler B (1996) The tumor necrosis factor ligand and receptor families. N Engl J Med 334:1717–1725

Bowie A, Kiss-Toth E, Symons JA, Smith GL, Dower SK, O'Neill LA (2000) A46R and A52R from vaccinia virus are antagonists of host IL-1 and toll- like receptor signaling. Proc Natl Acad Sci USA 97:10162–10167

Chaudhary PM, Ferguson C, Nguyen V, Nguyen O, Massa HF, Eby M, Jasmin A, Trask BJ, Hood L, Nelson PS (1998) Cloning and characterization of two Toll/Interleukin-1 receptor-like genes TIL3 and TIL4: evidence for a multi-gene receptor family in humans. Blood 91:4020–4027

Fitzgerald KA, Palsson-McDermott EM, Bowie AG, Jefferies CA, Mansell AS, Brady G, Brint E, Dunne A, Gray P, Harte MT, McMurray D, Smith DE, Sims JE, Bird TA, O'Neill LA (2001) Mal (MyD88-adapter-like) is required for Toll-like receptor-4 signal transduction. Nature 413:78–83

Fukui A, Inoue N, Matsumoto M, Nomura M, Yamada K, Matsuda Y, Toyoshima K, Seya T (2001) Molecular cloning and functional characterization of chicken toll-like receptors: a single chicken toll covers multiple molecular patterns. J Biol Chem 276:47143–47149

Gery I, Gershon RK, Waksman BH (1972) Potentiation of the T-lymphocyte response to mitogens. I. The responding cell. J Exp Med 136:128–142

Hemmi H, Takeuchi O, Kawai T, Kaisho T, Sato S, Sanjo H, Matsumoto M, Hoshino K, Wagner H, Takeda K, Akira S (2000) A Toll-like receptor recognizes bacterial DNA. Nature 408:740–745

Horng T, Barton GM, Medzhitov R (2001) TIRAP: an adapter molecule in the Toll signaling pathway. Nat Immunol 2:835–841
Inamori K, Koori K, Mishima C, Muta T, Kawabata S (2000) A horseshoe crab receptor structurally related to Drosophila Toll. J Endotoxin Res 6:397–399
Iwami KI, Matsuguchi T, Masuda A, Kikuchi T, Musikacharoen T, Yoshikai Y (2000) Cutting edge: naturally occurring soluble form of mouse Toll-like receptor 4 inhibits lipopolysaccharide signaling. J Immunol 165:6682–6686
Kadowaki N, Ho S, Antonenko S, Malefyt RW, Kastelein RA, Bazan F, Liu YJ (2001) Subsets of human dendritic cell precursors express different toll-like receptors and respond to different microbial antigens. J Exp Med 194:863–869
Krug A, Towarowski A, Britsch S, Rothenfusser S, Hornung V, Bals R, Giese T, Engelmann H, Endres S, Krieg AM, Hartmann G (2001) Toll-like receptor expression reveals CpG DNA as a unique microbial stimulus for plasmacytoid dendritic cells which synergizes with CD40 ligand to induce high amounts of IL-12. Eur J Immunol 31:3026–3037
Kumar S, Hedges SB (1998) A molecular timescale for vertebrate evolution. Nature 392:917–920
Lachman L, Hacker MP, Handschumacher RE (1977) Partial purification of human lymphocyte activating factor (LAF) by ultrafiltration and electrophoretic techniques. J Immunol 119:2019–2023
Li W-H (1997) Rates and patterns of nucleotide substitution. In: Molecular Evolution. Sunderland, MA, Sinauer Associates, Inc., pp 177–214
Libby P, Ordovas JM, Auger KR, Robbins AH, Birinyi LK, Dinarello CA (1986) Endotoxin and tumor necrosis factor induce interleukin-1 gene expression in adult human vascular endothelial cells. Am J Pathol 124:179–185
Matsuguchi T, Musikacharoen T, Ogawa T, Yoshikai Y (2000) Gene expressions of toll-like receptor 2, but not Toll-like receptor 4, is induced by LPS and inflammatory cytokines in mouse macrophages. J Immunol 165:5767–5772
Matsuguchi T, Takagi K, Musikacharoen T, Yoshikai Y (2000) Gene expressions of lipopolysaccharide receptors, Toll-like receptors 2 and 4, are differently regulated in mouse T lymphocytes. Blood 95:1378–1385
Medzhitov R, Preston-Hurlburt P, Janeway CA, Jr. (1997) A human homologue of the Drosophila Toll protein signals activation of adaptive immunity [see comments]. Nature 388:394–397
Musikacharoen T, Matsuguchi T, Kikuchi T, Yoshikai Y (2001) NF-κ B and STAT5 play important roles in the regulation of mouse Toll-like receptor 2 gene expression. J Immunol 166:4516–4524
Muzio M, Bosisio D, Polentarutti N, D'amico G, Stoppacciaro A, Mancinelli R, van't Veer C, Penton-Rol G, Ruco LP, Allavena P, Mantovani A (2000) Differential expression and regulation of toll-like receptors (TLR) in human leukocytes: selective expression of TLR3 in dendritic cells. J Immunol 164:5998–6004
Poltorak A, He X, Smirnova I, Liu M-Y, Van Huffel C, Du X, Birdwell D, Alejos E, Silva M, Galanos C, Freudenberg MA, Ricciardi-Castagnoli P, Layton B, Beutler B (1998) Defective LPS signaling in C3H/HeJ and C57BL/10ScCr mice: mutations in *Tlr4* gene. Science 282:2085–2088
Qureshi ST, Lariviere L, Leveque G, Clermont S, Moore KJ, Gros P, Malo D, Lipopolysaccharide, Inflammation, Positional c, Salmonella, Mice (1999) Endotoxin-tolerant mice have mutations in toll-like receptor 4 (Tlr4). J Exp Med 189:615–625
Qureshi ST, Larivière L, Leveque G, Clermont S, Moore KJ, Gros P, Malo D (1999) Endotoxin-tolerant mice have mutations in toll-like receptor 4 (Tlr4). J Exp Med 189:1519–1520
Rehli M, Poltorak A, Schwarzfischer L, Krause SW, Andreesen R, Beutler B (2000) PU.1 and interferon consensus sequence binding protein (ICSBP) regulate the myeloid expression of the human Toll-like receptor 4 gene. J Biol Chem 275:9773–9781
Rock FL, Hardiman G, Timans JC, Kastelein RA, Bazan JF (1998) A family of human receptors structurally related to Drosophila Toll. Proc Natl Acad Sci USA 95:588–593
Visintin A, Mazzoni A, Spitzer JH, Wyllie DH, Dower SK, Segal DM (2001) Regulation of Toll-like Receptors in human monocytes and dendritic cells. J Immunol 166:249–255
Wang T, Lafuse WP, Zwilling BS (2000) Regulation of toll-like receptor 2 expression by macrophages following Mycobacterium avium infection. J Immunol 165:6308–6313
Wang T, Lafuse WP, Zwilling BS (2001) NF-κB and Spl elements are necessary for maximal transcription of Toll-like receptor 2 induced by Mycobacterium avium. J Immunol 167:6924–6932
Yang R-B, Mark MR, Gray A, Huang A, Xie MH, Zhang M, Goddard A, Wood WI, Gurney AL, Godowski PJ (1998) Toll-like receptor-2 mediates lipopolysaccharide-induced cellular signalling. Nature 395:284–288

Plant Disease Resistance: Commonality and Novelty in Multicellular Innate Immunity

R. FLUHR and R.N. KAPLAN-LEVY

Pathogen avirulence genes encode for effector molecules that play a crucial role in the process of pathogen colonization of plant tissue. Successful host defense requires rapid and efficient detection of the pathogen avirulence factors. In the last few years, much progress has been made in delineating the plant molecular sentinels that participate in pathogen identification. Because this ability is genetic information that is 'hard-wired' into the genome, it is called 'innate immunity' and it draws its origins from a phylogenetically ancient form of immunity common to plants and animals. Conservation is shown in many of the functional molecular motifs of innate genes such as the Toll/interleukin 1 receptor domains, nucleotide binding domains and structures that contain leucine rich repeats. Novel plant molecular surveillance domains also include pathogen pattern recognition by coiled-coil domains and specialized kinases. The rapid evolution of plant innate immunity genes is readily detected in their sequence polymorphism, by their massive amplification and appearance in the genome in a clustered organization. By comparative biology of highly diverged innate immunity systems we can enhance our appreciation of the truly basic forces that have shaped its evolution in mutlicellular organisms.

Department of Plant Sciences, Weizmann Institute of Science Rehovot 76100, Israel,
e-mail: robertfluhr@weizmann.ac.il

1 Introduction

Pathogen avirulence genes encode for effector molecules that play a crucial role in the process of pathogen colonization of plant tissue. Successful host defense requires rapid and efficient detection of the pathogen avirulence factors. In the last few years, much progress has been made in delineating the plant molecular sentinels that participate in pathogen identification. Since this ability is genetic information that is 'hard-wired' into the genome, it is called 'innate immunity' and it draws its origins from a phylogenetically ancient form of immunity common to plants and animals. Conservation is shown in many of the functional molecular motifs of innate genes such as the Toll/interleukin (IL) 1 Receptor (TIR) domains, nucleotide binding domains and structures that contain leucine rich repeats (LRR). Novel plant molecular surveillance domains also include pathogen pattern recognition by coiled-coil (CC) domains and specialized kinases. The rapid evolution of plant innate immunity genes is readily detected in their sequence polymorphism, by their massive amplification and appearance in the genome in a clustered organization. By comparative biology of highly diverged innate immunity systems we can enhance our appreciation of the truly basic forces that have shaped its evolution in mutlicellular organisms.

2 Plant Defense Arsenals

A basic concept in immunity is the recognition of 'non-self' components as opposed to the components of 'self'. This is accomplished efficiently in plants by the presence of constitutive receptors called resistance genes (R-genes), which like tireless sentinels, alert the organism to activate a predetermined defense program. The activated plant cellular defense arsenal includes a large repertoire of general defense posturing that is not well differentiated for specificity of the response. Thus, a plant's response to diseases of viral, bacterial and fungal origins will lead to similar general defense patterns that include rapid oxidative bursts that can culminate in apoptopic-like cell death at the site of infection. When this effect is prominent, it is called the hypersensitive response and is a common, although not a necessary, feature of R-gene motivated responses (Cole et al. 2001). Other responses include induction of defensin-like molecules (Garcia-Olmedo et al. 2001; Kitajima and

SATO 1999), modification of the cell-wall (BRADLEY et al. 1992) and production of anti-microbial agents called phytoalexins (DIXON 2001). Importantly, the difference between a plant that can show resistance to a specific disease or its susceptible sibling that eventually succumbs is usually determined genetically by its pathogen recognition ability, i.e., the presence of the proper set of R-genes rather than the repertoire of its defense arsenal. Thus, a plant that is susceptible to one disease may be capable of mounting a highly successful normal defense pattern against another disease.

3 Plant R-Genes Are Defined Functionally by Specific Disease Resistance

The appreciation of how plant R-genes are detected at the genetic and molecular level is important for evaluating those special characteristics that differentiate them from their animal counterparts. Plant R-genes were first defined at a functional level, i.e., the ability to confer resistance to a specific disease. Thus they are present in cultivated crops – not by serendipity, but through active crop breeding programs that aim to continuously update and incorporate innate resistance genes from genetic pools that are available in closely related wild plant populations. The end result of these genetic crosses is introgression of a small area from the foreign genome into the cultivated crop. Once R-genes are characterized at a functional and genetic level, their molecular isolation can be accomplished by a variety of sophisticated genome positional cloning techniques or by transposon-mediated gene knockout (STASKAWICZ et al. 1995). In this manner, scores of individual R-genes have been isolated and shown to fall into six basic groups, each representing by inference supergene families of plant R-genes (Fig. 1).

4 The Molecular Building Blocks of Plant Innate Immunity Genes

Plant innate resistance genes are built up of putative recognition domains in combination with accessory motifs that participate either in membrane anchoring or in signal relay (Fig. 1). The largest family of plant R-genes of known function is made up of at least three discernable domains. These are called NB-LRR proteins and are composed of: an N-terminus consisting of a TIR or CC domain, an internal nucleotide binding (NB) domain and a C-terminal LRR domain (structure $B_{1/2}CA_1$; Fig. 1 and Sects. 5–7). Approximately 180 sequences of this type are annotated in the *Arabidopsis* genome. A close animal counterpart is emulated in NOD2, which consists of two N-terminal caspase activating recruitment domains (CARDs), and

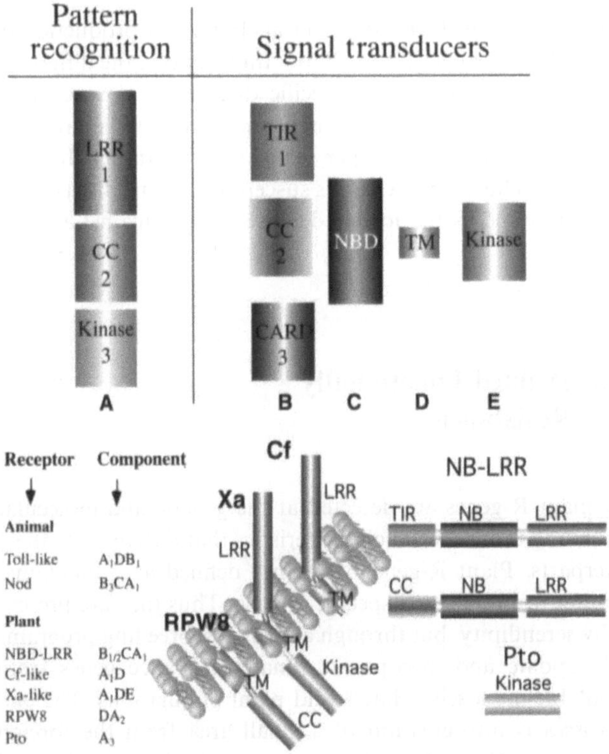

Fig. 1A–E. Pattern recognition receptors are built from domains common to plants and animals. In the upper panel, domains have been conceptually divided into five functional categories labeled A–E. **A** LRR, CC and a Kinase comprise domains that can function as pattern receptors. **B** TIR, CC and CARD are probably involved in signal transduction by homo- or hetero-dimerization with acceptor molecules. **C** NB domains common to NOD and plant TIR/CC-NB-LRR may serve a nucleotide-dependent switch function. **D** The transmembrane motif functions to anchor attached domains or participate in transmembrane signal transfer and may also contain endocytosis signals for signal attenuation as has been found in the tomato *Ve* Verticillium R-gene (Kawchuk et al. 2001). **E** The kinase domain probably participates in signal transduction. The *lower panel* shows domain combinations and cellular location of the main R-gene classes

a NB domain linked to multiple C-terminal LRRs (Ogura et al. 2001b). NOD2 was implicated in chronic inflammatory disorder of the gastrointestinal tract (Crohn's disease), and cells expressing NOD2 showed differential nuclear factor (NF)-κB activation when treated with bacterial lipopolysaccharides (LPSs) (Inohara et al. 2001; Ogura et al. 2001a). Furthermore, NOD proteins are cytosolic and a LPS binding activity could be specifically coimmunopurified from cytosolic extracts. Mammalian NOD proteins appear to be the animal counterparts of plant NB-LRR proteins. They function as cytosolic receptors for pathogen components derived from invading bacteria.

The additional classes of genes that have been implicated in plant innate immunity form a structurally diverse group (Fig. 1). They include: extracellular pattern recognition LRR domains that are either attached to the cell via a short membrane anchor (Cf-like), or to intracellular kinases (Xa-like or receptor-like kinases; RLK) via a transmembrane segment. These classes are represented by approximately 30 and 170 gene members, respectively, in the *Arabidopsis* genome. Animal innate immunity Toll-like receptor (TLR) proteins can be considered their closest counterpart. However, there are important differences. The Cf-like genes lack a TIR domain (structure A_1D; Fig. 1), though they may interact with a plethora of orphan independent TIR domains that are present in the *Arabidopsis* genome (Sect. 5.1). The Xa-like genes (A_1DE; Fig. 1) are remindful of condensed version of the signal transduction pathway of animal TLR through animal Pelle or IL-1 receptor associated kinase (IRAK) kinases. Indeed the Xa-like kinase domain belongs to the same superfamily of genes as Pelle suggesting their common origin and function (Shiu and Bleecker 2001).

Another large class of R-genes that were shown to contain pattern recognition ability are Pto-like kinases first isolated in tomato. Pto confers resistance to *Pseudomonas syringae* pv. *tomato*. Interestingly, the function of *Pto* requires the presence of *Prf*, a NB-LRR-type gene (Salmeron et al. 1996), yet it is the kinase that shows specific binding to its cognate avirulence factor (Sect. 8; Frederick et al. 1998). Homologue searches of the *Arabidopsis* genome detect about 50 members of this class and they are structurally similar to the Pelle superclass of kinases. Recently the discovery of *RPW8* defined yet another novel structural class containing a membrane anchor and a C-terminal CC domains (DA_2). The presence of this gene confers broad spectrum resistance to mildew (Xiao et al. 2001). Animal homologues to this type of motif have yet to be described in the context of innate immunity.

Evolution appears to have freely juggled with the building blocks of innate immunity genes and the sentinel proteins are built up by the combinatorial use of the different components mentioned above. As shown in Fig. 1, each domain can be involved either in pattern recognition or take part in signal transduction. Thus, the LRR domain probably acts in pattern recognition in animals and plants and has been shown to associate with pathogen generated ligands (Sect. 7.1). However, evidence suggests that in plants additional domains including CC domains and kinases can participate in pattern recognition (Sects. 5.3 and 8). In addition, shuffling of TIR domains can modify specificity to pathogen pathovars (Sect. 5.2), suggesting that the functional sentinel complex may be a multi-unit complex.

5 Plant TIR Domains

The N-terminus of the first NB-LRR-type plant R-gene described was shown to contain a TIR-like homology (Whitham et al. 1994). This was of particular interest because it became known that TIR-containing genes were also essential for a particular pathway of fly innate immunity (Kimbrell and Beutler 2001; Lemaitre et al.

1996). In plant NB-LRR R-genes, the TIR motif appears at the N terminus whereas in animals the TIR domain is at the carboxyl end of a single-pass transmembrane receptor. Plant and animal TIR domains are highly diverged from each other and form two distinct classes separated by significant bootstrap evaluation (Fig. 2). As evidenced by the depth of the branches, significant diversification is an additional feature within both divisions. Despite the sequence divergence, more than ten areas of consensus motif homology can be delineated with the help of a domain detector that relies on multiple sequence alignment of representative family members of plant and animal TIR sequences (Fig. 3; Ponting et al. 2000). The ten conserved modular regions of known TIR as shown coincide approximately with the predicted five β-sheet and five α-helices regions (Xu et al. 2000). The structures makes up the core conserved sequence of the representative genes and contain at least 135 residues. The variance in size of up to 200 residue found in plant TIR may therefore reside in the loop regions of the TIR structure.

Structural analysis of the human TIR, TLR-1 and TLR-2, show that they can project a relatively conserved surface for binding to MyD88, a signal adapter molecule (Xu et al. 2000). MyD88 is essential for signal transduction of the immune response and contains a TIR domain together with a death factor domain (Medzhitov et al. 1998). The surface conservation detected in the TIR domain is important as it may explain how divergent TIR-containing genes can all signal through the MyD88 adapter molecule. Interestingly, a recently found MyD88 homologue shows preference for interaction with TLR-4 and IRAK2 facilitating specificity in the innate immune response (Fitzgerald et al. 2001; Horng et al. 2001). Searches of plant databases have yet to reveal a plant MyD88 homologue, yet, the finding that the *Arabidopsis EDS1* mutation compromises the articulation of distinct TIR-NB-LRR disease R-genes argues for signal funneling through a common intermediate (Aarts et al. 1998b).

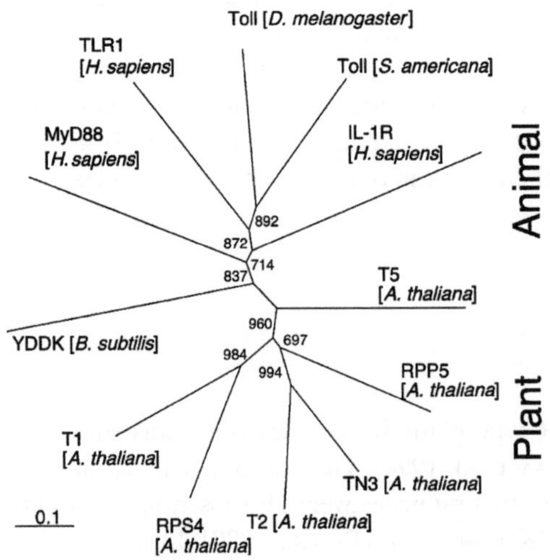

Fig. 2. Radial phylogenetic tree of representative animal and plants TIR domains. Multiply aligned representative sequences were compared by the neighbor-joining method (Saitou and Nei 1987). The tree is rooted using the distantly related bacterial YDDK sequence of *Bacillus subtilis* defined as an outgroup. Plant genes T1, T2 and T5 contain a TIR domain only (R.N. Kaplan-Levy and R. Fluhr, unpublished results). TN3 and RPS4 and RPP5 are TIR domains juxtaposed to NB domains alone or NB-LRR sequences of functional R-genes. Animal genes shown are IL-1R, TLR, Toll and MyD88 sequences

```
                 ←Box1→                                     ←-----Box2-----→
                 ←-------------TIR1-------------→
AT2g16870        NYRVFASFHG---------------PDVRKSFLSHFRKQFICN----GITMFDDQKIVRS
CONSENSUS/80%    ..psalsatt.................h..thh..l.pt........tlpha.cp...tt
CONSENSUS/65%    passFlsaps...............tthppphlsplhcthpth....tlslahcs.h.ts
CONSENSUS/50%    paDlFlSFsu...............cslccsalscLtcshpst....plslFtDcphhuu
                    |-βA-|                |----αA-----|    |---βB--|
                      A                          B                    C

                                             ←-----TIR2------→
AT2g16870        QT-I---APSLTQGIRESKIS--IVILSKNYASSTWCLN--ELLEILKCRED-----IGQ
CONSENSUS/80%    t.........l.ptlppuphh..lhlho.phhps.hsh...Eh..hhp.h........t.
CONSENSUS/65%    th........lhpsIcpS+hh..lhllSpsahpSpaChp..Ehhthhcpt.t.....tth
CONSENSUS/50%    ssht.....pltcuIccS+hs..llllScsYssSsWCls..ELtphhcstpp.....ssp
                  |---αB---------|  |---βC--||------αC-------|
                                 D               E               F

                 ←--------TIR3---------→
AT2g16870        IVMTVFYGVDP----SDVRKQTGEFGTVFNKT----------------------------
CONSENSUS/80%    hll.lhht.......t.h.t..tpht.hh.....................    ......
CONSENSUS/65%    tllslhhps......pphpp.ssphtthhppp..................    ......
CONSENSUS/50%    plIPIhacsss....sclcppssclppshpcs..................    ......
                 |-βD|               |---αD---|
                     G                  H

                                                             ←Box3→
AT2g16870        --------------------CARRTEKER-------------RNWSQALNVVGNI
CONSENSUS/80%    ......................th.tt................hhtphhh.hs..
CONSENSUS/65%    ....................hhphspppt..............hWppshhthssh
CONSENSUS/50%    ....................htchscccp.............ppWccslpslsst
                                    |βE-|              |----αE-----|
                                       I                      J
```

Amino acid grouping

Class	Key	Residues
Alcohol	o	S,T
Aliphatic	l	I,L,V
Any	.	A,C,D,E,F,G,H,I,K,L,M,N,P,Q,R,S,T,V,W,Y
Aromatic	a	F,H,W,Y
Charged	c	D,E,H,K,R
Hydrophobic	h	A,C,F,G,H,I,K,L,M,R,T,V,W,Y
Negative	-	D,E
Polar	p	C,D,E,H,K,N,Q,R,S,T
Positive	+	H,K,R
Small	s	A,C,D,G,N,P,S,T,V
Tiny	u	A,G,S
Turnlike	t	A,C,D,E,G,H,K,N,Q,R,S,T

Fig. 3. Consensus animal and plant motifs in the TIR domain. The degree of homology is shown for all TIRs present in the database. Representative members of gene families were first compiled and the consensus terms for the domain was extracted by the Simple Modular Architecture Research Tool (SMART; http://smart.embl-heidelberg.de/). *Upper case letters* are single letter code for amino acids. The *lower case letters* are amino acid classes that represent particular residues as shown by the insert key. One plant gene representative, the *A thaliana* R gene homologue At2g16870, is shown as an example of a plant TIR. The positions of the five β-sheets and five α-helices were approximated according to TLR1 in the SMART multiple alignment (Xu et al. 2000). *Lines* shown are previously detected consensus sequences. *TIRs 1–3* refer to plant TIR motifs (Meyers et al. 1999) and *boxes 1–3* refer to previously described animal motifs (Mitcham et al. 2000). *Letters A–J* represent the ten conserved-motif area delineated in the SMART analysis

5.1 Phylogeny of TIR Domains

The low identity between amino acid sequence of plant and animal TLR1/2 TIR domains (less than 20%) obviates facile structural modeling. The plant TIR family as represented in the sequenced *Arabidopsis* genome can be divided into a few TIR phylogenetic groups (Fig. 2). One group includes more than 100 genes that are composed of TIR domains linked to a singular NB domain, or as part of NB-LRR sequences (representatives TN3, RPS4, and RPP5; Fig. 2). Another major group consists of more than 30 genes are either solitary TIR domains or TIR domains juxtaposed to non-NB domains of unknown functions (representatives T1, T2 and T5; Fig. 2; R.N. Kaplan-Levy and R. Fluhr, unpublished results). The analysis of the TIR phylogenetic tree supports and extends the clear division between animal and plant taxa (KIMBRELL and BEUTLER 2001). The animal taxa are further divided into at least two independent Toll-like and IL receptor-like families. The similarity, albeit restricted, between plant and animal TIR domains suggests a common unicellular ancestor. Indeed, sensitive SMART searches – which use reiterative sequence alignment together with broad definition of conserved polypeptide structural elements – of current databases have revealed distant TIR homologies in prokaryotes as well (SCHULTZ et al. 2000). One example is the *Bacillus subtilis* sequence YDDK of unknown function, that was used here for simulating a genetic outgroup as shown in Fig. 2. Despite their obviously ancient origin, TIR domains have not been detected in any of the cereal genomes (see Sects. 5.3 and 10). Possibly, their functionality in signal transduction has been replaced in cereal R-genes by the presence of a N-terminal CC domain (MEYERS et al. 1999; PAN et al. 2000b).

5.2 Mapping Essential TIR Elements by Mutation

The TIR domain is essential for tobacco N-gene mediated tobacco mosaic virus (TMV) resistance. Amino acids that affect *Drosophila* and human TIR-dependent signaling cause either partial or full loss of N-gene function (DINESH-KUMAR et al. 2000). For example, N-gene D46 falls in the equivalent position of a conserved aspartate on the human TLR2 surface. Whereas the mutation D46H completely eliminates function, the D46Y substitution [the normal composition for the human Type I IL-1 receptor (IL-1R)] had no effect. Interestingly, the partial loss-of-function mutations can act as dominant-negative mutations by promoting a systemic hypersensitive response when present together with the wild-type *N* allele.

TIR domains are thought to play a role in signal transduction. Unexpectedly, the plant TIR domain may also determine pathogen specificity. In flax rust resistance, swapping the TIR domains of the L6 and L7 R-genes switches their specificity (LUCK et al. 2000). This, together with evidence for diversifying selection in the TIR region of the flax rust R-genes, may indicate that pattern recognition in plants is sophisticated and that plant resistance sentinels operate as a multi facet complex.

5.3 TIR Domains Are Replaced by Coiled-Coil Domains in a Large Set of R-Genes

Coiled-coils are an oligomerization motif of helical structures that are made up bundles of two to five helices. A typical CC structure shows a heptad repeat. The first and fourth residue tend to be hydrophobic and the fifth and seventh residues are charged or polar. Coiled-coils are often capable of organization as autonomously folding units and can demonstrate interaction specificity (O'SHEA et al. 1992). A large subset of eudicot and cereal NB-LRR genes can be shown to contain a CC domain in their N-terminal region with more than 95% probability (PAN et al. 2000b). In this context, CC domains may serve the same function of adapter TIR-like motifs. An exceptional case is the R-gene RPW8, which is responsible for broad spectrum resistance to mildew. RPW8 has CC domains that appear at the C terminus of a predicted transmembrane or signal peptide domain (XIAO et al. 2001). In this case, the CC domain may be functionally analogous to the LRR domain and acts directly in pattern recognition of an avirulence factor that is common to many mildew pathovars. Alternatively, it may play an accessory role facilitating other pattern recognition molecules to function. RPW8 is encoded by a small gene family of five linked members but due to the nature of CC domains true assessment of their potential numbers in plant genomes will require sophisticated database mining as has been illustrated in yeast (NEWMAN et al. 2000).

6 NB Domain Function

The NB domain is characterized by several sequence motifs found in many ATP- and GTP-binding proteins (TRAUT 1994), including the Ras superfamily and the caspase pathway-related *Ced-4* and *Apaf-1* (LI et al. 1997). These genes in animals regulate the activity of proteases that can initiate apoptopic cell death. As defense mechanisms in plants include apoptopic-like hypersensitive responses, the appearance of gene homologies in this context is particularly intriguing. By analogy to Ras, the NB domains may serve as a switch function moderating the inter- or intra-molecular activity of the polypeptide. With the lack of actual data on binding of nucleotides to NB domains, alternative structural modeling of a subset of the NB region has been proposed and suggests homology to the receiver domain of HIS-ASP phosphoproteins typical of prokaryotic signaling pathways (RIGDEN et al. 2000). Additionally, the *Arabidopsis* RPP5 NB domain was shown to interact in the yeast two-hybrid system with RelA/SpoT gene homologues. In *Escherichia coli* these genes are involved in p/pppGpp stress effector molecule signaling (VAN DER BIEZEN et al. 2000). Whether this implies a similar function in plants is not known.

6.1 Mapping Essential NB Domain Elements by Mutation

Residue substitutions in the NB domain of the plant R-genes Prf, RPS2, and RPM1 fall into regions of the conserved block domains (PAN et al. 2000b). A systematic approach to elucidate regions critical for activity was carried out in the TMV resistance N-gene NB region. Mutations in the conserved nucleotide binding site, e.g., glycines or lysine (G216 and K222), led to loss of N-gene action but also interfered in a negative-dominant fashion when the mutant transgene was present in the normal N-gene background (Dinesh-Kumar et al. 2000). These results are consistent with that detected in other NB domain-containing proteins and point to mechanistic similarity. For example, mutations at position D301 lead to complete loss of function. The equivalent mutations of the conserved aspartate residue in CED4 disrupts the oligomerization process that is necessary for activating downstream caspases (Yang et al. 1998).

7 LRR Domain: Structural Considerations

LRR structures mediate protein–ligand interaction and recognition specificity (Buchanan and Gay 1996). What are the biophysical properties of the LRR domain that promoted its choice as a sentinel pattern receptor? A loosely conserved general LRR structure is present in all plant *R*-gene types described with the exception of *Pto* and *RPW8*. A rough structural platform for conceptualizing what a plant LRR may look like is based on the crystal structure of the porcine ribonuclease inhibitor (RI) (Kajava et al. 1995; Kobe and Deisenhofer 1994). The 15 LRRs of RI are composed of an inner 'solvent-exposed' surface rim comprised of β-sheets connected by an outer rim of α-helical segments. In RI the β-sheets are stabilized by a ladder of hydrogen bonds between conserved cysteines and asparagine side chains. The outer rim of α-helical segments force a curvature on the molecule so that it comes nearly full circle to form a horseshoe with a small opening. It is through this opening that the cognate ribonuclease interacts in a multipoint fashion with the solvent surface (Kobe and Deisenhofer 1995). Some plant LRR can contain many less repeats, for example 7 in the case of the sugar beet nematode resistance gene $Hs1^{pro1}$ R-gene (Cai et al. 1997), or many more than 15 LRR which is usually the case. Thus, a strict application of the RI model to plant LRR would seemingly force complete closure and limit the surface accessibility for potential protein–protein interaction. However, plant LRR are imperfect and inspection of potential α-helical segments that are interspersed between the β-sheets show the helical regions to be much less conserved (they are shorter or nonexistent). The LRR domain structure must therefore deviate from RI in a manner that would generate less constrained and more flexible molecules held together mainly by the hydrogen-bonds between the β-sheets.

7.1 Specificity in LRR Interactions

Specificity of the interaction between LRR domains and their potential pathogen-based ligands has been inferred in animal TIR-LRR by examining isolated TLR4 immunity genes from different species. TLR4 genes were found to impart to transformed cell lines their particular species-specific sensitivity when tested with a range of different pharmacological LPS structures (LIEN et al. 2000; POLTORAK et al. 2000). Similar results were obtained in transgenic plants in which resistance response was measured in organisms carrying modified LRR domains of specific R-genes. Thus, in tomato, the *Cf*-like R-gene confers resistance to infection by the biotrophic leaf mold pathogen *Cladosporium* and contains an extracellular LRR domain (structure A_1D; Fig. 1). By precise domain swapping, a number of sub-domains within the LRR could be shown to be essential for both *Cf-4* and *Cf-9* function (VAN DER HOORN et al. 2001; WULFF et al. 2001).

Direct evidence for interaction of plant LRR domains with pathogen avirulence factors is based on the finding that a single amino acid difference in the LRR domain distinguished susceptible and resistant alleles of the fungal *Magnapporthe grisea* rice NB-LRR type R-gene (BRYAN et al. 2000). In this case, the recombinant LRR domain of the resistant allele could be shown to interact directly with the pathogen avirulence factor in the yeast two-hybrid system while the susceptible allele displayed a much weaker interaction (JIA et al. 2000). The results of direct interaction assays between R-genes and their cognate avirulence factor was less straightforward in the case of the *Arabidopsis* NB-LRR type R-genes *RPS2* and *RPM1*. These R-genes confer disease resistance to the phytopathogenic bacterium *Pseudomonas syringae* carrying the avirulence factors avrRpt2 and avrB, respectively. Using transient expression in leaf mesophyll protoplasts, an in vivo complex containing both the RPS2 and AvrRpt2 proteins was detected by co-immunoprecipitation. However, an additional plant protein of approximately 75kDa was present, and unexpectedly RPS2 also formed a complex with a non-cognate avirulence factor, AvrB (LEISTER and KATAGIRI 2000) – a result that may indicate that plant R-genes are sequestered into higher order protein complexes. Similarly, TIR containing proteins IL-1R and TLR4, require accessory proteins such as IL-1RAcP and CD14, respectively, for activity (for a review see DAUN and FENTON 2000).

7.2 Dynamics of LRR Evolution

Solvent-exposed regions of the LRR domain are implicated in the porcine RI ribonuclease interaction. If RI represents a structural paradigm for R-gene avirulence factor interactions, the expectation is that solvent-exposed regions would undergo positive selection and evolve more rapidly than neighboring sequences. Detecting the pace of adaptive evolution is carried out by comparing the distribution of the ratio (K_a) of non-synonymous nucleotide substitutions per non-synonymous site to the (K_s) synonymous nucleotide substitution per synonymous site (LI 1993). K_s exceeds K_a when mutations that cause amino acid change are deleterious and will be eliminated from populations by 'purifying' selection. However, if the mutations are advantageous for the organism, natural selection favors sequence diversification and a K_a/K_s ratio that is larger than 1. When this method was applied to analysis of mammalian major histocompatibility complex class I and II genes, clear positive selection at antigen recognition sites could be detected (HUGHES and NEI 1988, 1989). Inspection of plant R-genes reveals a

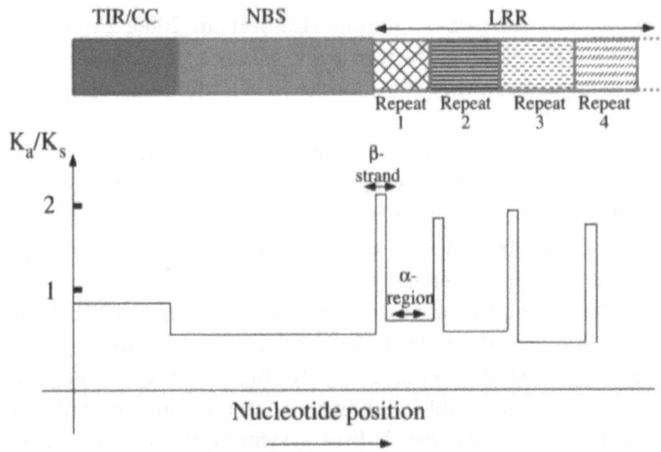

Fig. 4. Perceived ratio of non-synonymous/synonymous amino acid substitutions in NB-LRR haplotypes in tomato *Fusarium* R-genes. The schematic diagrams of the K_s and K_a values are presented for the sequenced paralogue members of the *I2 Fusarium oxysporum* cv. *lycopersici* R-gene family, I2C-2 vs. I2C-5 (ORI et al. 1997; SIMONS et al. 1998; SELA-BUURLAGE et al. 2001). The values were computed by Li's method (LI 1993). The solvent exposed LRR region includes consensus sequence xx(L)x(L)xx and is indicated by '*β-strand*'. The juxtaposed α-helical regions are indicated by '*α-region*'

similar phenomenon at select gene regions. Generally, the N-terminal CC structures as well as the NB domain show low average K_a/K_s ratio. In contrast, positive selection was found in the predicted β-strand region of the LRR domain of NB-LRR and LRR-kinase innate immunity genes (BOTELLA et al. 1998; MCDOWELL et al. 1998; MEYERS et al. 1998b; PARNISKE et al. 1997; WANG et al. 1998). The remarkable subtlety of this procedure is conceptualized in Fig. 4 with an example that compares K_a/K_s ratio along the NB-LRR-type resistance gene for the tomato *Fusarium oxysporum* R-gene *I2* (SELA-BUURLAGE et al. 2001). The evidence is consistent with rapid positive selection for polymorphism in the LRR domains of plant R-genes. Whether the evolutionary pace is facilitated by a propensity for local hypermutation rate, as has been detected for immunoglobin genes, remains to be seen (PAPAVASILIOU and SCHATZ 2000).

8 Pto, a Distant Homologue of Pelle, Interacts with an Avirulence Factor

The tomato *Pto* kinase is a plant R-gene displaying remarkable similarity to the IRAK and Pelle kinases that are active downstream of animal TIR-LRR-directed immune response (reviewed in COHN et al. 2001). Autophosphorylation competence was found to be obligatory for interaction of Pto and its cognate avirulence factor. Pto interacts with additional factors including Pti4–6 which are putative transcription factors (THARA et al. 1999). Pto requires for function the presence of the

Prf gene, a NB-LRR-type gene (SALMERON et al. 1996). There are approximately 50 *Pto* homologues in the *Arabidopsis* genome, a number which may argue for specific interaction of Pto-like kinases with a range of cognate NB-LRR. In a similar manner, the *Arabidopsis PBS1* gene encodes a putative kinase (that is distantly related to, but distinct from, the *Pto* family) that is necessary for RPS5 function but not the RPM1 or RPS2 NB-LRR resistance function (SWIDERSKI and INNES 2001).

Despite the rapid evolution of R-genes detected by phylogenetic analysis, it may be expected that the existence of persistent pathogens will help to maintain an ancient disease lineage. In the case of *Pto*, the wild tomato relatives *Lycopersicon pimpinellifolium* and *L. hirsutum* carry highly conserved homologues that interact with the Pto avirulence factor and confer resistance. However, the *L. esculentum* accessions examined contained no *Pto* orthologue. This suggests a common ancient origin for *Pto* in the two wild species and that Pto kinases similar to NB-LRR are part of dispensable pathways or have functionally redundant alleles.

Another type of serine-threonine kinases implicated in innate immunity appear as subdomains of RLK genes (A_1DE type structure; Fig. 1). RLKs belong to the RLK/Pelle superfamily of receptor-like kinases which comprises more than 600 members in *Arabidopsis* (SHIU and BLEECKER 2001). Some of these receptors are involved in development while others are involved in innate immunity pattern recognition (Sect. 12). The role of the kinase domain in RLK is probably restricted to signal transduction as was demonstrated by switching the extracellular LRR and transmembrane domains of the *Arabidopsis* brassinosteroid receptor kinase BRI1 with the serine/threonine kinase domain of the rice *XA21* resistance gene. The resultant transgene promoted brassinosteroid-dependent resistance type signaling (HE et al. 2000), implicating the LRR domain as responsible for ligand recognition and the kinase domain as responsible for pathway signaling.

9 The Meeting Place of Plant R-Genes and Their Cognate Pathogen Avirulence Factors

Although bacterial and fungal microorganisms grow along the surface of plant cells in the intercellular space, they undoubtedly must produce gene products that pass through or join the plant cell membrane. Obviously, in the case of viral infection the active part of the life cycle occurs intracellularly. For example, the presence of the *'N'* R-gene in tobacco plants limits the spread of TMV infection to small necrotic hypersensitive lesions. In this case, the pathogen avirulence factor detected by the plant is part of the helicase domain of the active viral replicase. This was evidenced by showing that a p50 recombinant fragment, whether in an enzymatically active state or not, was sufficient to induce N-gene dependent cell death (ERICKSON et al. 1999). The cognate plant R-gene (the *N*-gene) is thought to reside within the cell (WHITHAM et al. 1994). In contrast, a genetic system independent of the *N*-gene but with the capability of sensing the TMV viral coat protein was detected on the cell surface (ALLAN et al. 2001). The molecular nature of the surface gene product involved is unknown. Similarly, the tomato fungal leaf mold disease caused by *Cladosporium fulvum* is an example of a biological interaction in which the avirulence factor is detected outside of the plant cell (JONES et al. 1994). In this case, the relevant R-gene resides on the cell membrane and its cognate fungal avirulence factor was detected in the intercellular leaf space (DE WIT and JOOSTEN 1999; DEWIT 1997).

9.1 Cell–Cell Communication in Pathogen – Host Interactions

The majority of plant R-genes that have functional descriptions reside entirely within the cell. How then do pathogen effector molecules traverse the cell perimeter? In the case of bacteria, the type III secretion system has been implicated in promoting plant R-gene/avirulence factor interaction (LAHAYE and BONAS 2001; NIMCHUK et al. 2001; VAN DEN ACKERVEKEN et al. 1996). The type III protein secretion system can be likened to a molecular syringe that facilitates transfer of proteins from the bacterial pathogen to the host. An intriguing example is in the case of homologues of the *Yersinia* virulence effector *YopJ*. Homologues of *YopJ* are present in both plant and animal bacterial pathogens, as well as in plant symbionts. They can act as cysteine proteases and apparently disrupt mitogen-activated protein kinase and NF-κB dependent signaling in animal cells. Interestingly, the *Xanthomonas* plant pathogen *YopJ* homologue, *AvrBst*, was found to trigger localized R-gene dependent cell death (ORTH et al. 2000). Pathogen effector proteins not only cross the outer cell membrane but may be targeted to specific subcellular compartments. For example, the *AvrBs3* avirulence factor was found to contain nuclear targeting motifs (NLS) and has been show to interact with importin-α a molecule that mediates the passage of proteins through the nuclear pore complex. Disruption of the NLS motif eliminates avirulence factor recognition by the plant (SZUREK et al. 2001). In the case of fungal pathogens, the molecular mechanism that facilitates potential protein transfer remains to be elucidated.

A glimpse of the complex biology involved in plant receptor/fungal avirulence factor interactions is illustrated in rice blast disease caused by the fungus *Magnapporthe grisea*. The fungus encodes for a pre-propeptide (AVR-Pita) that has the features of a metalloprotease (ORBACH et al. 2000). The peptide with avirulence activity is apparently processed and is somehow transferred from the fungus *into* the plant cell. Only within the cell can it interact with its cognate cytoplasmic-localized NB-LRR indicating that plant–fungal interactions can emulate bacterial type III protein transfer mechanisms.

9.2 Alternative Cellular Targets for R-Genes

Simplistically, the avirulence factor is thought to interact with the R-gene in a receptor–ligand type of interaction as described in Section 7.1. However, the paucity of evidence confirming direct receptor–ligand interactions and the theoretical difficulty in understanding how a small number of innate immune genes could cope with a diverse pathogen world, has led to a more sophisticated proposition called the 'guard model'. This hypothesis assumes that pathogen effector molecules must interact with host proteins. The plant R-genes are meant to take the measure of the effector–pathogenicity target interaction (DANGL and JONES 2001). Thus, the R-gene may normally interact with the host target molecule, when this interaction is perturbed by the pathogen effector molecule the R-gene commences a pathogenesis response. In this manner, it would be necessary for the organism to maintain only a finite number of R-genes that are adapted to protect the likely targets of pathogenicity effectors. In this case, one may expect that R-genes for identical disease will display restricted taxonomic functionality. Evidence for this can be drawn from the fact that the *Bs2* pepper resistance gene and its cognate avirulence factor, *avrBs2*, actively produce hypersensitive response only in the closely related *Solanaceous* plant species but not in other plants (TAI et al. 1999). However, recent evidence for avirulence factor binding by LRR may indicate that both direct and indirect pathogen sensing strategies exist.

10 Rapid Evolution of Plant R-Genes and Loss of TIR Domains in Cereals

Plant R-genes belong to supergene families. The NB domains have served as a facile phylogenetic framework for isolating R-gene analogs. In contrast to the highly diverged TIR/CC and LRR domains that compose NB-LRR resistance homologues, all NB domains contain considerable conserved stretches of sequence homology. Their presence has ameliorated their use in phylogeny analysis in soybean, potato, rice, barley and *Arabidopsis* (AARTS et al. 1998a; KANAZIN et al. 1996; LEISTER et al. 1998; YU et al. 1996).

Phylogenetic analyses clearly show that NB domains can be divided into two distinct groups that contain particular subdomain signatures (MEYERS et al. 1999; PAN et al. 2000b). Both NB domain groups are present in eudicots in which one group is always found linked to resident TIR sequences in the N-terminal region and the other group is linked to resident CC in the N-terminal region. Surprisingly, this entire supergene group that contains TIR domains is notably absent from cereal (monocot) databases as these sequences cannot be amplified from cereal genomes (PAN et al. 2000b). Alternatively, in monocots this motif has evolved to be unrecognizable at the sequence level. The genetic loci *EDS1* and *NDR1* were shown to modulate either TIR-type or CC-type multiple R-gene action respectively in *Arabidopsis* (AARTS et al. 1998b), the lack of a whole group of R-gene analogs implies bifurcation of signal transduction pathways between these two great plant classes.

The ancient origin of TIR domains indicates that cereal genomes have lost or utterly corrupted their copies: the question is when did this occur? In addition to the detection of TIR in Metazoans, TIR-NB domains have been detected in Gynmnosperms suggesting that the Angiosperm progenitor plant contained TIR as well. The loss of TIR sequences in cereals (and perhaps all monocot genomes) must have occurred during monocot and dicot divergence over the last 170 million years. Mechanisms such as unequal crossing-over can account for elimination of small multigene families however no mechanism has been suggested for complete elimination of dispersed supergene families from genomes. It is therefore likely that before the loss of TIR domains only a few germline NB-LRR-type R-genes existed in the ancient angiosperm progenitor. A small multigene number would be more in-line with the lower multigene number of TIR-LRR detected in the Metazoan genomes.

10.1 Clustering of Plant Resistance Genes in the Genome and the Origin of New Resistance Alleles

NB-LRR sequences can reside in large extended arrays spanning millions of base pairs that can consist of dozens of contiguous R-gene homologues. In the *Arabidopsis* genome about 77% of the detected NB domain sequences appear as clusters of more than one gene (http://niblrrs.ucdavis.edu). Other plant species show similar patterns of clustering at particular NB-LRR sequences that are known to be involved in disease resistance, for example: 15 NB domain sequences in the M locus of flax (ANDERSON et al. 1997); eight in the Xa21 locus of rice (SONG et al. 1997); seven at the tomato *I2* locus (SIMONS et al. 1998) and 24 in the *Dm3* cluster of lettuce (MEYERS et al. 1998a). Similarly, on a genome-wide basis, NB domain analogs have been shown to cluster in soybean (KANAZIN et al. 1996; YU et al. 1996), cereal genomes (LEISTER et al. 1998) and *Solanaceae* (GEBHARDT and VALKONEN 2001; GRUBE et al. 2000; PAN et al. 2000a).

Is clustering the result of recent or ancient adaptive duplications? Is there a selective advantage to resistance fitness in maintaining clusters? Clustering is thought to be one mechanism of generating rapid sequence diversity. Juxtaposing sequences would initially be the source of new alleles and variance due to unequal recombination and gene conversion events. However, a thorough evaluation of different size clusters in the *Arabidopsis* genome reveals that K_a/K_s ratios are more or less similar, i.e., they are high regardless of the complexity of the cluster. This observation is true as well for single isolated R-genes for which paralogues in the genome could be found, indicating that additional forces facilitate rapid evolution. Yet, it is of interest that clusters which contain the most highly related paralogues show the highest K_a/K_s ratio, a finding which suggests even more rapid selection in the clustered newly duplicated alleles (BERGELSON et al. 2001). Thus while isolated R-genes can still maintain a high adaptive rate, in some cases clustered genes appear to be evolving even more rapidly. Definitive answers as to what selective advantages plant genomes have in maintaining dynamic gene clusters will be resolved when more evolutionary data gathered through population genetics becomes available.

10.2 R-Gene Synteny and Conservation of Function

An underlying assumption is that higher plants have inherited their gene order and content from a common ancestor. Given the rapid evolution of R-genes due to environmental pressure it is of interest to compare the status of these genes relative to general genome architecture. Remarkably, despite the rapid intrinsic evolution of R-genes a great deal of cross-generic genome synteny can be detected. Global conservation of synteny despite speciation has been observed in *Solanaceae* together with specific R-gene analog loss (PAN et al. 2000a). Similar loss, but with much more frequent non-syntenic map positions for cereal R-genes and their homologues may indicate more rapid sequence divergence in the monocots (LEISTER et al. 1998). The fact that cereals completely lack the more conserved TIR-LRR type R-genes is consistent with their more rapid global rate of R-gene evolution. It is important to keep in mind that even when clusters survive speciation there is no reason to expect functional conservation. Examination of species within the *Solanaceae* family readily detected conserved structural positioning but with completely different disease specificity. For example, resistance to *Fusarium* in tomato corresponds to resistance to TMV in pepper on chromosome 11, and the tomato nematode *Hero* R-gene corresponds to a potato *Phytophthora* resistance loci on chromosome 4 (GEBHARDT and VALKONEN 2001; GRUBE et al. 2000). However, more closely related species exhibit a high degree of functional conservation. This question was approached by examining directly the disease resistance potential of two related genomes in one uniform genetic background. To accom plish this a series of introgression lines carrying the complete *L. pennellii* genome in the *L. esculentum* background was examined for *Fusarium* resistance. Many independent loci were detected with varying quantitative differences between them, and two out of six loci show common evolutionary origins by appearing in the same chromosomal location in more than one species (SELA-BUURLAGE et al. 2001).

10.3 Population Dynamics of R-Gene Evolution

The adaptive nature of pathogen–plant interaction argues for continual selective turnover of alleles. The 'arms-race' hypothesis predicts that as a result of recent selective sweeps in a population, both resistance and sensitive alleles should be young and of related sequence (STAHL et al. 1999). Polymorphism of plant R-genes as a source for new resistance variation can be assessed by comparing orthologous genes in different ecological accessions of the same species and in different species. In this manner, 17 accessions of *A. thaliana* were sequenced to explore the diversity present in resistance and susceptibility alleles of *Rps2*. The alleles which confer resistance to *P. syringae* were found to be more conserved than the susceptibility alleles (Caicedo et al. 1999). Resistance and sensitive alleles (27 in number) at the *Rpm1* locus showed significant sequence divergence between them, a finding that prompted the authors to reject the 'arms-race' hypothesis (Stahl et al. 1999). Maintenance of polymorphism by the population requires striking a balance between the necessity for resistance in the presence of the pathogen and the cost of resistance when the particular pathogen is absent. The changes in resistance allele frequency support the concept of continual cyclic 'trench warfare' between the pathogen and its host.

11 The Genetic Expense of Environmental Sensing

Animal innate immunity consists of a limited number of pattern recognition genes that assist and prime the more complex acquired immunity response. It complements the acquired immune system, which consists of a huge array of limited populations of somatically produced recombinant proteins. As acquired immunity attains complexity during the organism's somatic stage it offers the benefit of complete 'non-self' coverage without a huge penalty in stored genetic information. Hence, innate systems are inherently more costly in their 'information content'. However, the developmental strategy of plants, with their rigid cellular structures and absence of a circulatory system, would seem to preclude potential evolutionary development of the sophisticated acquired pattern recognition systems. Instead, plant sentinels are required to be completely cell autonomous. Thus, every plant cell or at least a subset of cells targeted by the pathogen must be able to play the role of sentinel and alert its own cellular machinery and probably also that of more distant cells to the presence of a pathogen. There is no reason to surmise that plants recognize fewer potential pathogens than their animal counterparts. How, then, can an organism that relies solely on innate immunity cope? The animal innate immunity repertoire of pattern recognition is increased by using dynamic hetero- and homo-dimerization of TLRs (Ozinsky et al. 2000). It is not known whether such interactions exist for plant R-genes. Plants have apparently met the challenge of pathogen detection by choosing as their sentinels a versatile group of pattern recognition molecules, by greatly amplifying their numbers and by positioning them in the genome in a manner that facilitates their rapid evolution.

Simplifying the vast complexity of potential non-self ligands to common pattern recognition elements is one means of coping with variety in pathogen recognition. Many pathogens must breach the polymers in the plant cell wall in order to colonize the plant. The resultant release of a common set of oligosaccharides acts as a signal for defense gene activation (Darvill et al. 1992). In other cases, pattern receptors can be sensitive to pathogen-specific marker polymers such as chitin or β (1–3) glucans. The recent characterization of the *Arabidopsis FLS2* innate immunity gene is the case closest to classical pattern recognition by a LRR receptor kinase (Gomez-Gomez and Boller 2000). FLS2 is an A_1DE type innate receptor (Fig. 1) and can detect the pattern of a conserved short amino acid domain in bacteria flagellin (FELIX et al. 1999).

Most of the pathogen avirulence factors characterized bear little resemblance to each other at the linear sequence level. Surprisingly, despite the lack of apparent homology, a single R-gene can recognize more than one avirulence factor. *RPM1* has been implicated in recognition of avrRpt2 and avrB (GRANT et al. 1995) and multi-recognition of avirulence factors is probably true for *Mi*, which confers resistance to both nematodes and aphids in tomato (ROSSI et al. 1998). It is possible that three-dimensional structural commonalties exist between the avirulence factors that have been overlooked. Alternatively, the 'guard model' (Sect. 9.2) would explain multiple specificity as being due to interaction of the different avirulence factors with the same complex of cellular machinery that is guarded by a particular R-gene.

11.1 The Smell of Danger as a Paradigm for Plant Innate Immunity Systems

Animal and plant innate immunity systems show clear conservation in evolutionary origin, yet the massive amplification of R-gene family members presents a special strategic opportunity for environmental sensing. A useful paradigm for plant–pathogen sensing is animal receptors of odors. In higher vertebrates fewer than 1,000 odor receptors are sufficient to 'sense' a great spectrum of aroma. How does the nose cope? LANCET and coworkers described a probability model that predicts the repertoire size of smell receptors necessary for proper representation of receptors with a specified affinity (LANCET et al. 1993). They conclude that at the expense of reduced receptor affinity the complexity of the repertoire of potential interactions (compared to that obtained with antibodies) can be reduced. By setting an affinity value of $10^5 M^{-1}$, a number that has been measured experimentally for some odorants, a repertoire of smell receptors 300–1,000 in number would suffice for finding with high probability at least one receptor with the required specificity.

Similarly, if the average affinities of R-genes for their ligands were modest, smaller populations would suffice. This may explain the experimental difficulty in actually measuring R-gene–avirulence interactions above the background level. The outcome of lower association constants is the expectation that one receptor may recognize multiple ligands, a finding that is true for odorants (MALNIC et al. 1999) as well as R-genes. The comparison of innate plant immunity to odor sensing has its limitations. In animal sensory systems, the additional higher order neural input integrates information from different receptors and yields combinatorial output (MALNIC et al. 1999). Plant response to a variety of pathogen input appears to be more uniform and different pathogen inputs, whether of bacterial or fungal origin, lead to similar transcriptome patterns (MALECK et al. 2000).

Clearly the application of powerful genetic tools to discern features of plant innate immunity has drawn a complex, and as yet far from complete, picture of the way in which these molecular sentinels function. The tantalizing commonalties between plant and animal sentinel domains and components intimates to their early origin. While innate immunity appears to be a specialized phenomenon characteristic of multicellular organisms, the Metazoa and Viridiplantae separated at the unicellular stage. Implying that these functions, perhaps in a profoundly altered state, were and probably are present in single cell organisms as well. One has only to look. Undoubtedly, the insight garnered from comparative biology of highly divergent organisms will be central to the appreciation of the biological forces that dictate the overall strategy of innate immunity.

References

Aarts MGM, Hekkert BT, Holub EB, Beynon JL, Stiekema WJ, Pereira A (1998a) Identification of R-gene homologous DNA fragments genetically linked to disease resistance loci in *Arabidopsis thaliana*. Molecular Plant-Microbe Interactions 11:251–258

Aarts N, Metz M, Holub E, Staskawicz BJ, Daniels MJ, Parker JE (1998b) Different requirements for EDS1 and NDR1 by disease resistance genes define at least two R gene-mediated signaling pathways in *Arabidopsis*. Proc Natl Acad Sci USA 95:10306–10311

Allan AC, Lapidot M, Culver JN, Fluhr R (2001) An early tobacco mosaic virus-induced oxidative burst in tobacco indicates extracellular perception of the virus coat protein. Plant Physiol 126:97–108

Anderson PA, Lawrence GJ, Morrish BC, Ayliffe MA, Finnegan EJ, Ellis JG (1997) Inactivation of the flax rust resistance gene M associated with loss of a repeated unit within the leucine-rich repeat coding region. Plant Cell 9:641–651

Bergelson J, Kreitman M, Stahl EA, Tian D (2001) Evolutionary dynamics of plant R-genes. Science 292:2281–2285

Botella MA, Parker JE, Frost LN, Bittner-Eddy PD, Beynon JL, Daniels MJ, Holub EB, Jones JDG (1998) Three genes of the *Arabidopsis RPP1* complex resistance locus recognize distinct *Peronospora parasitica* avirulence determinants. Plant Cell 10:1847–1860

Bradley DJ, Kjellbom P, Lamb CJ (1992) Elicitor- and wound-induced oxidative cross-linking of a proline-rich plant cell wall protein: a novel, rapid defense response. Cell 70:21–30

Bryan GT, Wu KS, Farrall L, Jia Y, Hershey HP, McAdams SA, Faulk KN, Donaldson GK, Tarchini R, Valent B (2000) A single amino acid difference distinguishes resistant and susceptible alleles of the rice blast resistance gene Pi-ta. Plant Cell 12:2033–2046

Buchanan SG, Gay NJ (1996) Structural and functional diversity in the leucine-rich repeat family of proteins. Prog Biophys Mol Biol 65:1–44

Cai D, Kleine M, Kifle S, Harloff HJ, Sandal NN, Marcker KA, Klein-Lankhorst RM, Salentijn EM, Lange W, Stiekema WJ, Wyss U, Grundler FM, Jung C (1997) Positional cloning of a gene for nematode resistance in sugar beet. Science 275:832–834

Caicedo AL, Schaal BA, Kunkel BN (1999) Diversity and molecular evolution of the RPS2 resistance gene in *Arabidopsis thaliana*. Proc Natl Acad Sci USA 96:302–306

Cohn J, Sessa G, Martin GB (2001) Innate immunity in plants. Curr Opin Immunol 13:55–62

Cole AB, Kiraly L, Ross K, Schoelz JE (2001) Uncoupling resistance from cell death in the hypersensitive response of *Nicotiana* species to cauliflower mosaic virus infection. Mol Plant Microbe Interact 14:31–41

Dangl JL, Jones JDG (2001) Plant pathogens and integrated defence responses to infection. Nature 411:826–833

Darvill A, Augur C, Bergmann C, Carlson RW, Cheong JJ, Eberhard S, Hahn MG, Lo VM, Marfa V, Meyer B, Mohnen D, Oneill MA, Spiro MD, Vanhalbeek H, York WS, Albersheim P (1992) Oligosaccharins – oligosaccharides that regulate growth, development and defense responses in plants. Glycobiology 2:181–198

Daun JM, Fenton MJ (2000) Interleukin-1/Toll receptor family members: receptor structure and signal transduction pathways. J Interferon Cytokine Res 20:843–855 de Wit PJ, Joosten MH (1999) Avirulence and resistance genes in the *Cladosporium fulvum*-tomato interaction. Curr Opin Microbiol 2:368–73

DeWit P (1997) Pathogen avirulence and plant resistance: a key role for recognition. Trends in Plant Science 2:452–458

Dinesh-Kumar SP, Tham WH, Baker BJ (2000) Structure-function analysis of the tobacco mosaic virus resistance gene N. Proc Natl Acad Sci USA 97:14789–14794

Dixon RA (2001) Natural products and plant disease resistance. Nature 411:843–7

Erickson FL, Holzberg S, Calderon-Urrea A, Handley V, Axtell M, Corr C, Baker B (1999) The helicase domain of the TMV replicase proteins induces the N-mediated defence response in tobacco. Plant J 18:67–75

Felix G, Duran JD, Volko S, Boller T (1999) Plants have a sensitive perception system for the most conserved domain of bacterial flagellin. Plant J 18:265–276

Fitzgerald KA, Palsson-McDermott EM, Bowie AG, Jefferies CA, Mansell AS, Brady G, Brint E, Dunne A, Gray P, Harte MT, McMurray D, Smith DE, Sims JE, Bird TA, O'Neill LA (2001) Mal (MyD88-adapter-like) is required for Toll-like receptor-4 signal transduction. Nature 413:78–83

Frederick RD, Thilmony RL, Sessa G, Martin GB (1998) Recognition specificity for the bacterial avirulence protein *AvrPto* is determined by Thr-204 in the activation loop of the tomato Pto kinase. Mol Cell 2:241–245

Garcia-Olmedo F, Rodriguez-Palenzuela P, Molina A, Alamillo JM, Lopez-Solanilla E, Berrocal-Lobo M, Poza-Carrion C (2001) Antibiotic activities of peptides, hydrogen peroxide and peroxynitrite in plant defence. FEBS Lett 498:219–22

Gebhardt C, Valkonen JPT (2001) Organization of genes controlling disease resistance in the potato genome. Annu Rev Phytopathol 39:79–102

Gomez-Gomez L, Boller T (2000) FLS2: an LRR receptor-like kinase involved in the perception of the bacterial elicitor flagellin in *Arabidopsis*. Mol Cell 5:1003–1011

Grant MR, Godiard L, Straube E, Ashfield T, Lewald J, Sattler A, Innes RW, Dangl JL (1995) Structure of the *Arabidopsis RPM1* gene enabling dual specificity disease resistance. Science 269:843–846

Grube RC, Radwanski ER, Jahn M (2000) Comparative genetics of disease resistance within the solanaceae. Genetics 155:873–887

He Z, Wang ZY, Li J, Zhu Q, Lamb C, Ronald P, Chory J (2000) Perception of brassinosteroids by the extracellular domain of the receptor kinase BRI1. Science 288:2360–2363

Horng T, Barton GM, Medzhitov R (2001) TIRAP: an adapter molecule in the Toll signaling pathway. Nat Immunol 2:835–841

Hughes AL, Nei M (1988) Pattern of nucleotide substitution at major histocompatibility complex class I loci reveals overdominant selection. Nature 335:167–170

Hughes AL, Nei M (1989) Nucleotide substitution at major histocompatibility complex class II loci: evidence for overdominant selection. Proc Natl Acad Sci USA 86:958–962

Inohara N, Ogura Y, Chen FF, Muto A, Nunez G (2001) Human Nod1 confers responsiveness to bacterial lipopolysaccharides. J Biol Chem 276:2551–2554

Jia Y, McAdams SA, Bryan GT, Hershey HP, Valent B (2000) Direct interaction of resistance gene and avirulence gene products confers rice blast resistance. EMBO J 19:4004–4014

Jones DA, Thomas CM, Hammond-Kosack KE, Balint-Kurti PJ, Jones JDG (1994) Isolation of the tomato *Cf-9* gene for resistance to *Cladosporium fulvum* by tranposon tagging. Science 266:789–793

Kajava AV, Vassart G, Wodak SJ (1995) Modeling of the three-dimensional structure of proteins with the typical leucine-rich repeats. Structure 3:867–877

Kanazin V, Marek LF, Shoemaker RC (1996) Resistance gene analogs are conserved and clustered in soybean. Proc Natl Acad Sci USA 93:11746–11750

Kawchuk LM, Hachey J, Lynch DR, Kulcsar F, van Rooijen G, Waterer DR, Robertson A, Kokko E, Byers R, Howard RJ, Fischer R, Prufer D (2001) Tomato Ve disease resistance genes encode cell surface-like receptors. Proc Natl Acad Sci USA 98:6511–6515

Kimbrell DA, Beutler B (2001) The evolution and genetics of innate immunity. Nat Rev Genet 2:256–267

Kitajima S, Sato F (1999) Plant pathogenesis-related proteins: molecular mechanisms of gene expression and protein function. J Biochem (Tokyo) 125:1–8

Kobe B, Deisenhofer J (1994) The leucine-rich repeat: a versatile binding motif. Trends Biochem Sci 19:415–421

Kobe B, Deisenhofer J (1995) A structural basis of the interactions between leucine-rich repeats and protein ligands. Nature 374:183–186

Lahaye T, Bonas U (2001) Molecular secrets of bacterial type III effector proteins. Trends Plant Sci 6:479–485

Lancet D, Sadovsky E, Seidemann E (1993) Probability model for molecular recognition in biological receptor repertoires: significance to the olfactory system. Proc Natl Acad Sci USA 90:3715–3719

Leister D, Kurth J, Laurie DA, Yano M, Sasaki T, Devos K, Graner A, Schulze-Lefert P (1998) Rapid reorganization of resistance gene homologues in cereal genomes. Proc Natl Acad Sci USA 95:370–375

Leister RT, Katagiri F (2000) A resistance gene product of the nucleotide binding site – leucine rich repeats class can form a complex with bacterial avirulence proteins in vivo. Plant J 22:345–354

Lemaitre B, Nicolas E, Michaut L, Reichhart JM, Hoffmann JA (1996) The dorsoventral regulatory gene cassette spatzle/Toll/cactus controls the potent antifungal response in Drosophila adults. J Biol Chem 271:5777–5783

Li P, Nijhawan D, Budihardjo I, Srinivasula SM, Ahmad M, Alnemri ES, Wang X (1997) Cytochrome c and dATP-dependent formation of Apaf-1/caspase-9 complex initiates an apoptotic protease cascade. Cell 91:479–489

Li WH (1993) Unbiased estimation of the rates of synonymous and nonsynonymous substitution. J Mol Evol 36:96–99

Lien E, Means TK, Heine H, Yoshimura A, Kusumoto S, Fukase K, Fenton MJ, Oikawa M, Qureshi N, Monks B, Finberg RW, Ingalls RR, Golenbock DT (2000) Toll-like receptor 4 imparts ligand-specific recognition of bacterial lipopolysaccharide. J Clin Invest 105:497–504

Luck JE, Lawrence GJ, Dodds PN, Shepherd KW, Ellis JG (2000) Regions outside of the leucine-rich repeats of flax rust resistance proteins play a role in specificity determination. Plant Cell 12:1367–1377

Maleck K, Levine A, Eulgem T, Morgan A, Schmid J, Lawton KA, Dangl JL, Dietrich RA (2000) The transcriptome of *Arabidopsis thaliana* during systemic acquired resistance. Nat Genet 26:403–410

Malnic B, Hirono J, Sato T, Buck LB (1999) Combinatorial receptor codes for odors. Cell 96:713–723

McDowell JM, Dhandaydham M, Long TA, Aarts MGM, Goff S, Holub EB, Dangl JL (1998) Intragenic recombination and diversifying selection contribute to the evolution of downy mildew resistance at the RPP8 locus of arabidopsis. Plant Cell 10:1861–1874

Medzhitov R, Preston-Hurlburt P, Kopp E, Stadlen A, Chen C, Ghosh S, Janeway CA, Jr. (1998) MyD88 is an adaptor protein in the hToll/IL-1 receptor family signaling pathways. Mol Cell 2: 253–258

Meyers BC, Chin DB, Shen KA, Sivaramakrishnan S, Lavelle DO, Zhang Z, Michelmore RW (1998a) The major resistance gene cluster in lettuce is highly duplicated and spans several megabases. Plant Cell 10:1817–1832

Meyers BC, Dickerman AW, Michelmore RW, Sivaramakrishnan S, Sobral BW, Young ND (1999) Plant disease resistance genes encode members of an ancient and diverse protein family within the nucleotide-binding superfamily. Plant J 20:317–332

Meyers BC, Shen KA, Rohani P, Gaut BS, Michelmore RW (1998b) Receptor-like genes in the major resistance locus of lettuce are subject to divergent selection. Plant Cell 10:1833–1846

Mitcham JL, Parnet P, Bonnert TP, Garka KE, Gerhart MJ, Slack JL, Gayle MA, Dower SK, Sims JE (2000) T1/ST2 signaling establishes it as a member of an expanding interleukin- 1 receptor family. J Biol Chem 275:4670–4678

Newman JR, Wolf E, Kim PS (2000) A computationally directed screen identifying interacting coiled coils from Saccharomyces cerevisiae. Proc Natl Acad Sci USA 97:13203–13208

Nimchuk Z, Rohmer L, Chang JH, Dangl JL (2001) Knowing the dancer from the dance: R-gene products and their interactions with other proteins from host and pathogen. Curr Opin Plant Biol 4:288–94

O'Shea EK, Rutkowski R, Kim PS (1992) Mechanism of specificity in the Fos-Jun oncoprotein heterodimer. Cell 68:699–708

Ogura Y, Bonen DK, Inohara N, Nicolae DL, Chen FF, Ramos R, Britton H, Moran T, Karaliuskas R, Duerr RH, Achkar JP, Brant SR, Bayless TM, Kirschner BS, Hanauer SB, Nunez G, Cho JH (2001a) A frameshift mutation in NOD2 associated with susceptibility to Crohn's disease. Nature 411:603–606

Ogura Y, Inohara N, Benito A, Chen FF, Yamaoka S, Nunez G (2001b) Nod2, a Nod1/Apaf-1 family member that is restricted to monocytes and activates NF-κB. J Biol Chem 276:4812–4818

Orbach MJ, Farrall L, Sweigard JA, Chumley FG, Valent B (2000) A telomeric avirulence gene determines efficacy for the rice blast resistance gene Pi-ta. Plant Cell 12:2019–2032

Ori N, Eshed Y, Paran I, Presting G, Aviv D, Tanksley S, Zamir D, Fluhr R (1997) The I2C family from the wilt disease resistance locus I2 belongs to the nucleotide binding, leucine-rich repeat superfamily of plant resistance genes. Plant Cell 9:521–532

Orth K, Xu Z, Mudgett MB, Bao ZQ, Palmer LE, Bliska JB, Mangel WF, Staskawicz B, Dixon JE (2000) Disruption of signaling by *Yersinia* effector Yop J, a ubiquitin-like protein protease. Science 290:1594–1597

Ozinsky A, Underhill DM, Fontenot JD, Hajjar AM, Smith KD, Wilson CB, Schroeder L, Aderem A (2000) The repertoire for pattern recognition of pathogens by the innate immune system is defined by cooperation between toll-like receptors. Proc Natl Acad Sci USA 97:13766–13771

Pan Q, Liu YS, Budai-Hadrian O, Sela M, Carmel-Goren L, Zamir D, Fluhr R (2000a) Comparative genetics of nucleotide binding site-leucine rich repeat resistance gene homologues in the genomes of two dicotyledons: tomato and arabidopsis. Genetics 155:309–322

Pan Q, Wendel J, Fluhr R (2000b) Divergent evolution of plant NBS-LRR resistance gene homologues in dicot and cereal genomes. J Mol Evol 50:203–213

Papavasiliou FN, Schatz DG (2000) Cell-cycle-regulated DNA double-stranded breaks in somatic hypermutation of immunoglobulin genes. Nature 408:216–221

Parniske M, Hammond-Kosack K, Golstein C, Thomas CM, Jones DA, Harrison K, Wulff BB, Jones JD (1997) Novel disease resistance specificities result from sequence exchange between tandemly repeated genes at the *Cf-4/9* locus of tomato. Cell 91:821–832

Poltorak A, Ricciardi-Castagnoli P, Citterio S, Beutler B (2000) Physical contact between lipopolysaccharide and Toll-like receptor 4 revealed by genetic complementation. Proc Natl Acad Sci USA 97:2163–2167

Ponting CP, Schultz J, Milpetz F, Bork P (2000) SMART: identification and annotation of domains from signalling and extracellular protein sequences. Nucleic Acids Res 28:231–234

Rigden DJ, Mello LV, Bertioli DJ (2000) Structural modeling of a plant disease resistance gene product domain. Proteins 41:133–143

Rossi M, Goggin FL, Milligan SB, Kaloshian I, Ullman DE, Williamson VM (1998) The nematode resistance gene *Mi* of tomato confers resistance against the potato aphid. Proc Nat Acad Sci USA 95:9750–9754

Saitou N, Nei M (1987) The neighbor-joining method: a new method for reconstructing phylogenetic trees. Mol Biol Evol 4:406–425

Salmeron JM, Oldroyd GE, Rommens CM, Scofield SR, Kim HS, Lavelle DT, Dahlbeck D, Staskawicz BJ (1996) Tomato Prf is a member of the leucine-rich repeat class of plant disease resistance genes and lies embedded within the *Pto* kinase gene cluster. Cell 86:123–133

Schultz J, Copley RR, Doerks T, Ponting CP, Bork P (2000) SMART: a web-based tool for the study of genetically mobile domains. Nucleic Acids Res 28:231–234

Sela-Buurlage MB, Budai-Hadrian O, Pan Q, Carmel-Goren L, Vunsch R, Zamir D, Fluhr R (2001) Genome-wide dissection of *Fusarium* resistance in tomato reveals multiple complex loci. Mol Genet Genomics 265:1104–1111

Shiu SH, Bleecker AB (2001) Receptor-like kinases from *Arabidopsis* form a monophyletic gene family related to animal receptor kinases. Proc Natl Acad Sci USA 98:10763–10768

Simons G, Groenendijk J, Wijbrandi J, Reijans M, Groenen J, Diergaarde P, van der Lee T, Bleeker M, Onstenk J, de Both M, Haring M, Mes J, Cornelissen B, Zabeau M, Vos P (1998) Dissection of the fusarium I2 gene cluster in tomato reveals six homologs and one active gene copy. Plant Cell 10: 1055–1068

Song WY, Pi LY, Wang GL, Gardner J, Holsten T, Ronald PC (1997) Evolution of the rice *Xa21* disease resistance gene family. Plant Cell 9:1279–1287

Stahl EA, Dwyer G, Mauricio R, Kreitman M, Bergelson J (1999) Dynamics of disease resistance polymorphism at the Rpm1 locus of *Arabidopsis*. Nature 400:667–671

Staskawicz BJ, Ausubel FM, Baker BJ, Ellis JG, Jones JD (1995) Molecular genetics of plant disease resistance. Science 268:661–667

Swiderski MR, Innes RW (2001) The *Arabidopsis PBS1* resistance gene encodes a member of a novel protein kinase subfamily. Plant J 26:101–112

Szurek B, Marois E, Bonas U, van den Ackerveken G (2001) Eukaryotic features of the *Xanthomonas* type III effector AvrBs3: protein domains involved in transcriptional activation and the interaction with nuclear import receptors from pepper. Plant J 26:523–534

Tai TH, Dahlbeck D, Clark ET, Gajiwala P, Pasion R, Whalen MC, Stall RE, Staskawicz BJ (1999) Expression of the *Bs2* pepper gene confers resistance to bacterial spot disease in tomato. Proc Natl Acad Sci USA 96:14153–14158

Thara VK, Tang X, Gu YQ, Martin GB, Zhou JM (1999) *Pseudomonas syringae* pv *tomato* induces the expression of tomato EREBP-like genes pti4 and pti5 independent of ethylene, salicylate and jasmonate. Plant J 20:475–483

Traut TW (1994) The functions and consensus motifs of nine types of peptide segments that form different types of nucleotide-binding sites. Eur J Biochem 222:9–19

van den Ackerveken G, Marois E, Bonas U (1996) Recognition of the bacterial avirulence protein AvrBs3 occurs inside the host plant cell. Cell 87:1307–1316

van der Biezen EA, Sun J, Coleman MJ, Bibb MJ, Jones JD (2000) *Arabidopsis* RelA/SpoT homologs implicate (p)ppGpp in plant signaling. Proc Natl Acad Sci USA 97:3747–3752

van der Hoorn RA, Roth R, De Wit PJ (2001) Identification of distinct specificity determinants in resistance protein cf-4 allows construction of a cf-9 mutant that confers recognition of avirulence protein avr4. Plant Cell 13:273–285

Wang GL, Ruan DL, Song WY, Sideris S, Chen L, Pi LY, Zhang S, Zhang Z, Fauquet C, Gaut BS, Whalen MC, Ronald PC (1998) Xa21D encodes a receptor-like molecule with a leucine-rich repeat domain that determines race-specific recognition and is subject to adaptive evolution. Plant Cell 10:765–779

Whitham S, Dinesh-Kumar SP, Choi D, Hehl R, Corr C, Baker B (1994) The product of the tobacco mosaic virus resistance gene *N*: similarity to Toll and the interleukin-1 receptor. Cell 78:1101–1115

Wulff BB, Thomas CM, Smoker M, Grant M, Jones JD (2001) domain swapping and gene shuffling identify sequences required for induction of an avr-dependent hypersensitive response by the tomato Cf-4 and Cf-9 proteins. Plant Cell 13:255–272

Xiao S, Ellwood S, Calis O, Patrick E, Li T, Coleman M, Turner JG (2001) Broad-spectrum mildew resistance in *Arabidopsis thaliana* mediated by RPW8. Science 291:118–120

Xu Y, Tao X, Shen B, Horng T, Medzhitov R, Manley JL, Tong L (2000) Structural basis for signal transduction by the Toll/interleukin-1 receptor domains. Nature 408:111–115

Yang X, Chang HY, Baltimore D (1998) Essential role of CED-4 oligomerization in CED-3 activation and apoptosis. Science 281:1355–1357

Yu YG, Buss GR, Saghai Maroof MA (1996) Isolation of a superfamily of candidate disease resistance genes in soybean based on a conserved nucleotide-binding site. Proc Natl Acad Sci USA 93: 11751–11756

Signal Transduction Pathways Activated by the IL-1 Receptor/Toll-Like Receptor Superfamily

L.A.J. O'NEILL

Toll-like receptors (TLRs) are an important point of first contact between host and microbe, and once activated generate signals which culminate in the induction of genes important for host defence. TLRs respond to different microbial products, and the signalling pathways activated are very similar to that generated by the pro-inflammatory cytokine interleukin-1 (IL-1). This is because the Type I IL-1 receptor and TLRs are highly homologous in their cytosolic portions, possessing a Toll/IL-1 receptor (TIR) domain. Signals triggered include the important transcription factor NF-κB and two MAP kinases, p38 and Jun N-terminal kinase. Receptor-proximal proteins involved include the adapter MyD88, IRAK, IRAK-2, Tollip, TRAF6 and TAK-1. These latter two proteins need to be ubiquitinated in order to be active. Differences between signals generated by TLRs are emerging, with TLR-4 signalling requiring an additional adapter termed MyD88-adapter-like (Mal), which may regulate the expression of genes specific for the response required to eliminate infection by Gram-negative bacteria. Future studies on TLR signalling may reveal hitherto unsuspected specificities in the innate immune response to infection.

1 Introduction

A key feature of the innate immune response concerns the recognition of microbe-derived molecules by Toll-like receptors (TLRs). Upon recognition, the receptors trigger signal transduction pathways that lead to the increased expression of genes involved in host defence. An important theme that is starting to emerge is that the genes induced tailor the immune response to the pathogen, giving a degree of

---

Department of Biochemistry, Trinity College Dublin, Ireland, *e-mail: laoneill@tcd.ie*

specificity to innate immunity hitherto unsuspected. The basis for this specificity will lie not only in the receptors recognizing the microbial products, but also in the signal transduction pathways activated.

There are currently ten known human TLRs, and each recognizes a different microbial product. TLR-2 is involved in the recognition of peptidoglycan from Gram-positive bacteria (TAKEUCHI et al. 1999; SCHWANDER et al. 1999) bacterial lipoproteins (ALIPRANTIS et al. 1999; BRIGHTBILL et al. 1999) and zymosan from yeast (UNDERHILL et al. 1999), TLR-3 recognizes viral RNA (ALEXOPOULOU et al. 2001), TLR-4 recognizes lipopolysaccharide (LPS) (POLTORAK et al. 1998; QURESHI et al. 1999; HOSHINO et al. 1999), TLR-5 recognizes flagellin (HAYASHI et al. 2001), TLR-6 recognizes mycobacterial lipopeptide (TAKEUCHI et al. 2001) and finally TLR-9 recognizes CpG DNA (HEMMI et al. 2000). The role of the remaining TLRs is still not known, although both TLR-1 and TLR-6 can form dimers with TLR-2 (OZINSKY et al. 2000; TAKEUCHI et al. 2001), and it is likely that TLRs will homodimerize, e.g., in the case of TLR-4 OZINSKY et al. 2000, or heterodimerize (OZINSKY et al. 2000; TAKEUCHI et al. 2001) during microbial recognition. An important question is how receptor activation leads to changes in the expression of specific genes. Many investigations have focussed on pathways already shown to be activated by the pro-inflammatory cytokine interleukin (IL)-1 because the Type I IL-1 receptor (IL-1RI) and TLRs show a high level of sequence similarity in their cytosolic regions important for signalling (GAY and KEITH 1991; O'NEILL and GREENE 1998). These signals include the transcription factor nuclear factor (NF)-κB, which is an important regulator of a large number of genes involved in inflammation and immunity, and stress-activated protein kinases such as p38 MAP kinase and Jun N-terminal kinase (JNK), both of which have been shown to regulate the stability of mRNAs which encode inflammatory proteins. However, it is likely that TLRs will also generate specific signals. This will lead to both common sets of genes being induced and subsets specific for each receptor, which will be required for elimination of the provoking infectious agent.

2 IL-1 and Toll Signalling

Two fields of investigation acted as important precursors for the field of TLRs. The first of these concerned IL-1, an important pro-inflammatory cytokine with a wide range of activities, while the second concerned a protein in *Drosophila melanogaster* called Toll. The effects of IL-1 are due to induction of genes such as those encoding COX-2, iNOS, adhesion molecules, acute phase proteins, many other cytokines and chemokines and matrix metalloproteinases. This last response implicates IL-1 as a key participant in the joint erosion which occurs in diseases such as rheumatoid arthritis, and IL-1 remains an important therapeutic target in such diseases. Much work focussed on IL-1 signal transduction and the gene encoding the receptor for IL-1, termed the Type I IL-1 receptor (IL-1RI), was cloned in 1988 (SIMS et al. 1988). Its sequence gave no clues whatsoever as to how it might signal. For example, by that time it was known that many growth factor receptors had tyrosine kinase activity. Even though IL-1 has growth factor-like effects in some cell types, its receptor was not a tyrosine kinase. In fact, at the time of its identification IL-1RI was unique amongst receptors in its cytosolic domain: although similar to many immunomodulatory proteins, it possessed immunoglobulin domains extracellularly. In 1991 however, an important sequence similarity was noted, when GAY and KEITH reported that the cytosolic domain of IL-1RI had a high degree of similarity with a protein from *Drosophila melanogaster* termed Toll (GAY and KEITH 1991). The *Toll* gene was first described in 1985 by Anderson and Nusslein-Volhard (ANDERSON et al. 1985), and was named Toll because of the 'weirdness' of the embryos which develop when Toll is mutated – Toll being the German for 'weird'. Toll's role was in the establishment of dorso-ventral polarity in the developing *Drosophila* embryo (ANDERSON et al. 1985). Why a protein in the fly involved in development would be homologous to a mammalian protein involved in inflammation was not clear. However, the transcription factor which Toll activates is termed Dorsal, which is held in check by another protein, Cactus (ROTH et al. 1989; KIDD 1992). Dorsal, like the subunits which make up NF-κB in mammals, is a member of the Rel family of transcription factors, whilst Cactus is homologous to the NF-κB inhibitor I-κB. It therefore appeared that similar signalling machines were operating – one involved in inflammation, the other involved in dorso-ventral polarity. Two other proteins were described on the Toll pathway – Tube and Pelle (LETSOU et al. 1991; SHELTON and WASSERMAN 1993). Pelle is a homologue of the IL-1 receptor-associated kinase (IRAK) (CAO et al. 1996a), which is important for both IL-1 and TLR signalling, further strengthening the similarity between the two pathways. An important breakthrough was then made in 1996, when Hoffmann and colleagues reported that in the adult fly, the Toll pathway is required for defence against fungi (LEMAITRE et al. 1996). Here, Toll activated another Rel family member, termed Dif, again in a pathway involving Tube, Pelle and Cactus (LEMAITRE et al. 1996; WU and ANDERSON 1998). Dif regulates the expression of anti-fungal genes indicating that the innate response in flies is pathogen-specific. Another pathway in the fly, also involving a Rel family member termed Relish and an as-yet unidentified receptor, is triggered by bacteria and culminates in the induction of anti-bacterial peptides (HEDENGREN et al. 1999). Importantly, Toll does not appear to be activated directly by fungi. Instead the fungi lead to the production of a ligand for Toll via a serine protease caspase, termed Spätzle (LEVASHINA et al. 1999), which also is required for Toll function in development. This may be different in mammals, where there is evidence indicating that LPS is the ligand for TLR-4 (DA SILVA CORREIA et al. 2001), although recently we have found evidence for serine protease involvement in TLR-4 activation by LPS (MANSELL et al. 2002).

3 A New Receptor Superfamily Defined: the TIR Domain

Based on the homology between IL-1RI and Toll, it was possible to define a new domain, termed the Toll/IL-1 receptor domain (TIR domain) and there are 20 proteins with TIR domains so far identified in the human genome (O'NEILL 2000). Many other TIR-domain- containing proteins have been described in other species, most notably in plants, where they also have a role in host defence (O'NEILL 2000). There is also a Toll in the nematode, and again its role is in defence (PUJOL et al. 2001). It is involved in the sensing of bacteria, the signal it generates being very unusual – it provokes the nematode to move away from pathogenic bacteria. Because the TIR domain is present in so many species, it can be concluded that it evolved early in the evolution of life, occurring in the common unicellular ancestor of plants and animals and possibly even in prokaryotes (O'NEILL 2000). The TIR domain may therefore be one of the first domains involved in host defence to have evolved.

The TIR domain is approximately 200 amino acids long with three key patches of conservation (called boxes 1–3; O'NEILL 2000), which occur in all members, from plants to mammals. Some of the conserved amino acids had been shown to be important for signalling in site-directed mutagenesis studies (e.g. a key tryptophan in box-3; HEGUY et al. 1992). The structure of the TIR domain in TLR-1 and TLR-2 has been solved (XU et al. 2000) and is shown for TLR-2 in Fig. 1A. It reveals that one of the highly conserved regions in the TIR domain, Box 2, protrudes from the surface forming what is called the BB loop, and is likely to act as a docking site for the signalling adapter MyD88 (see below). The proline residue in Box 2 occurs at the tip of the BB loop and is worthy of mention because it is this amino acid that is mutated in the *TLR-4* gene from the LPS hyporesponsive C3H/HeJ mouse, the discovery of which led to the identification of TLR-4 as the receptor for LPS (POLTORAK et al. 1998; QURESHI et al. 1999; HOSHINO et al. 1999). The mutation converts the proline to a histidine, and although this does not alter the structure of the TIR domain (which was somewhat surprising), it introduces a charge which may alter the nature of the interaction with MyD88. This is supported by the observation that when there is a valine in this position (as in Toll itself) signalling is not affected. If the valine in Toll is substituted by a proline, signalling is normal. However, if the valine is substituted by a histidine, arginine or glutamic acid, signalling is impaired (XU et al. 2000). A final point is that the structures of the TIR domains in TLR-1 and TLR-2 although broadly similar show subtle differences (XU et al. 2000). This might lead to differences in signal transduction processes activated by each receptor.

In humans the TIR superfamily can be split into three subgroups, as shown in Fig. 1B. The first of these is the IL-1R subgroup, which is defined by extracellular immunoglobulin domains and includes IL-1RI and it accessory protein IL-1RAcP, IL-18R and its accessory protein, and T1/ST2. Several members of this subgroup are still orphans, although with the recent description of six additional IL-1-like molecules in the human genome (DUNN et al. 2001), it is

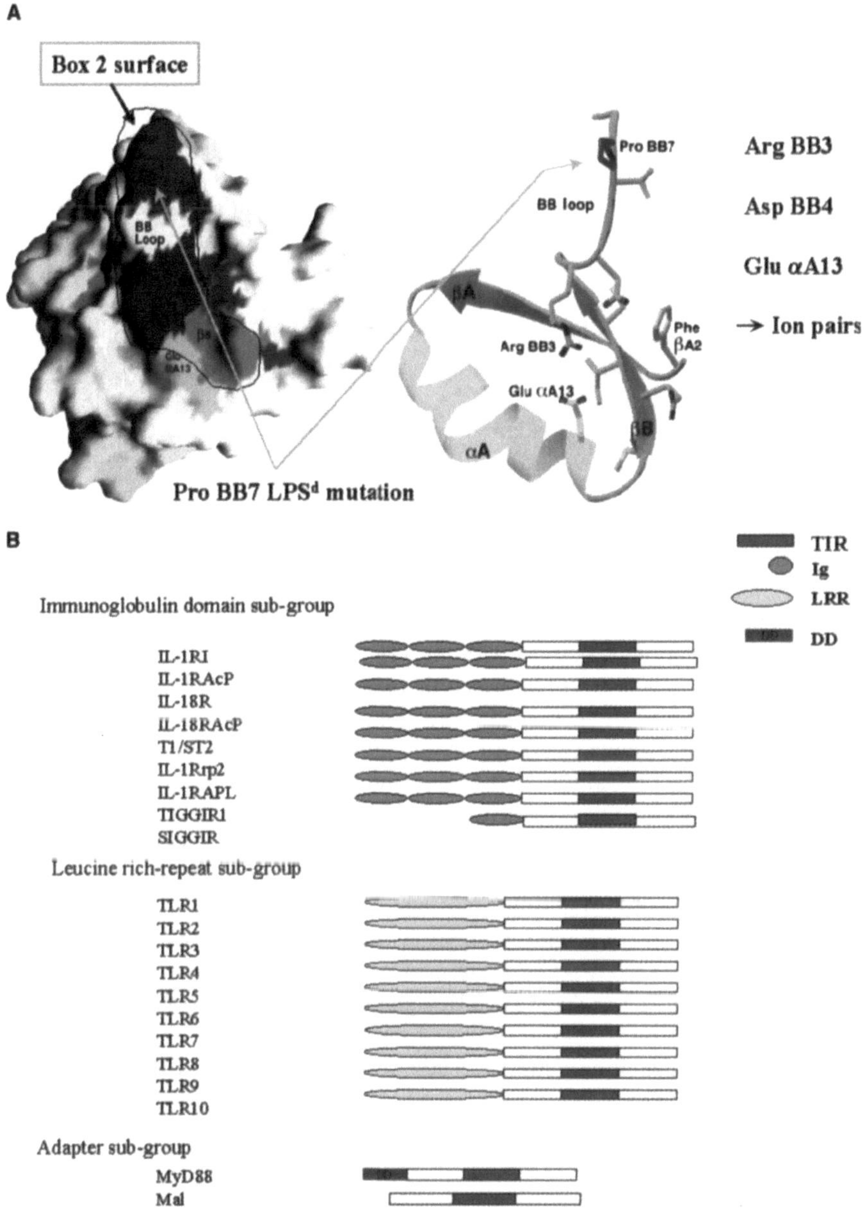

Fig. 1A,B. TIR domain: structure and proteins in the human genome predicted to have TIR domains. **A** The molecular surface of the TIR domain of TLR-2 is shown on the *left*, with the residues contributing to the BB loop shown on the *right* (From Xu et al. 2000). The conserved Box 2 of the TIR domain, present in a wide range of proteins from plants to animals, forms a loop termed the BB loop, which is on the surface of the TIR domain as indicated. The proline in Box 2 occurs at the top of this loop, the structural integrity of which is maintained by ion pairs, as shown on the *right hand side*. It is likely that the BB loop is the point of contact between the receptor TIR domain and the TIR domain in the adapter MyD88. **B** Proteins with TIR domains in the human genome are shown. The superfamily splits into three subgroups: those with immunoglobulin domains extracellularly, those with leucine-rich repeats extracellularly and the cytosolic adapter subgroup. See text for further details. *TIR*, Toll IL-1 receptor domain; *Ig*, immunoglobulin domain; *LRR*, leucine-rich repeat domain; *DD*, death domain

likely that their ligands will be identified soon. The second subgroup contains the TLRs themselves, so-called because of their extracellular and intracellular homology with Toll (which comprises leucine-rich repeats and a TIR domain respectively). The first of these to be reported was what became called TLR-1 (SIMS and DOWER 1994). This was followed by a key paper describing hToll, subsequently renamed TLR-4 (MEDZHITOV et al. 1997). A constitutively active form of TLR-4 was shown to activate NF-κB, induce cytokines such as tumour necrosis factor (TNF) and co-stimulatory molecules such as B7, implicating a human Toll in innate immunity, and providing evidence for evolutionary conservation in innate immunity between species. There is now a total of 10 TLRs, with functions as described above.

The third subgroup is the adapter subgroup which comprises two members, MyD88 (WESCHE et al. 1997; MEDZHITOV et al. 1998; BURNS et al. 1998) and MyD88 adapter-like (Mal) (FITZGERALD et al. 2001) (which has also been named TIRAP (Toll/IL-1 receptor domain containing adapter protein; HORNG et al. 2001), both of which are cytosolic. MyD88 was first described in 1990 as a protein induced during myeloid differentiation (LORD et al. 1990). It has a TIR domain and a death domain. It was then shown that MyD88 is required for IL-1 signalling, becoming associated in a homophilic manner with the TIR domain on IL-1RI (WESCHE et al. 1997; BURNS et al. 1998; MEDZHITOV et al. 1998) (see below). MyD88 has subsequently been shown to be involved in signalling by several TLRs (GEWIRTZ et al. 2001; YANG et al. 1999; TAKEUCHI et al. 2000; ADACHI et al. 1998; KAWAI et al. 1999). The role of Mal is in TLR-4 signalling (FITZGERALD et al. 2001; HORNG et al. 2001). It does not appear to be involved in signalling by IL-1, IL-18, TLR-2 or TLR-9 (FITZGERALD et al. 2001; HORNG et al. 2001), and is therefore the first protein to be described which is specific for signal transduction by a TIR superfamily member. It is discussed in greater detail below.

Finally, there are two viral proteins which show homology with the TIR domain (BOWIE et al. 2000). They are termed A46R and A52R and are in *Vaccinia virus*. They appear to act as decoys in TIR signalling, blocking NF-κB activation by IL-1 and TLR-4. Most recently we have found that A52R appears to sequester IRAK-2 from the signalling pathway. These results imply that TLRs are required for anti-viral host defence.

4 IL-1 Signal Transduction

IL-1RI is the most studied of all the TIR domain-containing receptors and the signalling pathway it activates has informed the analysis of other receptors in the superfamily. It had been noted that IL-1 and LPS generate very similar signals [e.g. NF-κB and p38 mitogen-activated protein (MAP) kinase activation], which was sometimes interpreted as being due to LPS inducing IL-1 production, which would then act in an autocrine manner. We now know, however, that the similarities are due to the TIR domain being in IL-1RI and TLR-4. All of the proteins described for IL-1 signalling have been implicated in LPS signal transduction, and most have been implicated in IL-18R, TLR-2, TLR-5 and TLR-9 signalling, thereby defining the set of signals shared by these receptors.

The signal transduction pathway for IL-1 is shown in Fig. 2. IL-1 signal transduction starts with IL-1 binding to IL-1RI, which is probably in a pre-formed complex with IL-1 receptor accessory protein IL-1RAcP. IL-1RI then undergoes a conformational change, where the third immunoglobulin domain moves towards the ligand (SCHREUDER et al. 1997). This conformational change is probably transmitted to IL-1RAcP. The alterations in extracellular conformation are likely to alter the interaction between the TIR domains of IL-1RI and IL-1RAcP,

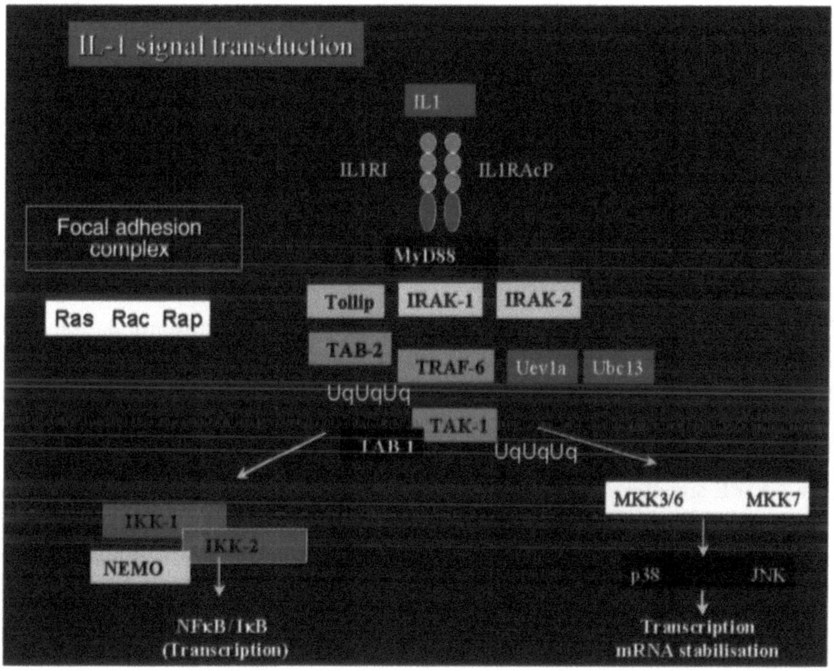

Fig. 2. IL-1 signal transduction. IL-1 signalling is initiated upon binding of IL-1 to IL-1RI. A signalling complex between IL-1RI and IL-1RAcP is then formed which recruits the adapter MyD88. The proteins IRAK and IRAK-2 are then recruited which are complexed to Tollip and possibly TAB-2. TRAF6 is then recruited in association with the ubiquitinating enzymes Ubc13 and Uev1A. TRAF6 becomes ubiquitinated and forms a complex with the kinase TAK-1, which is regulated by TAB-2 and TAB-1. TAK-1 then becomes ubiquitinated and is then active. TAK-1 activates IKK-2 leading to NF-κB activation, and also activates upstream kinases for p38 and JNK. These signals results in increased transcription of target genes and also increased mRNA stability in the case of p38 and JNK. A role for the FAC and the low molecular weight G proteins Ras, Rac and Rap has also been indicated. Similar cascades are likely to operate for TLRs. See text for further details. *IL-1RI*, IL-1 Type I receptor; *IL-1RAcP*, IL-1 receptor accessory protein; *IRAK*, IL-1 receptor-associated kinase; *TRAF6*, TNF receptor-associated factor-6; *TAK-1*, TGFβ-activated kinase; *IKK*, I-κB kinase; *NEMO*, NF-κB essential modulator; *MKK*, Map kinase kinase

allowing them to recruit MyD88 (Wesche et al. 1997; Medzhitov et al. 1998; Burns et al. 1998). It is likely that a similar scenario occurs for TLRs, which as stated above, probably all act as dimers. The recruitment of MyD88 to the receptor complex then leads to the recruitment of IRAK and IRAK-2 (or in myeloid cells IRAK-M) (Cao et al. 1996a; Muzio et al. 1997; Wesche et al. 1999). Both of these proteins have death domains, and so are probably recruited via a death domain–death domain interaction with MyD88. IRAK is a serine/threonine kinase, although its substrates are poorly defined. During signalling, IRAK becomes hyperphosphorylated and is subsequently degraded by the proteasome (Yamin and Miller 1997). The kinase responsible for IRAK phosphorylation is unknown. It is unlikely to be autophosphorylation, as a kinase-dead form of IRAK still becomes hyperphosphorylated, and in fact can function like wild-type IRAK. This has been shown in complementation studies with IRAK-deficient cells, which can be restored upon transfection with IRAK or kinase-dead IRAK (Li et al. 1999). Neither IRAK-2 nor IRAK-M have any kinase activity (Muzio et al. 1997; Wesche et al. 1999), and IRAK-2 can compliment for IRAK (Wesche et al. 1999). The IRAKs are therefore likely to act as adapters.

The kinase responsible for IRAK phosphorylation may be recruited by the region in MyD88 between the death domain and the TIR domain. This has been suggested from a study describing a splice variant of MyD88 lacking this inter-region, which acts as an inhibitor of IL-1 signalling (Janssens et al. 2002). The splice variant can still associate with IRAK but IRAK does not become phosphorylated.

Apart from tagging IRAK for subsequent degradation, phosphorylation may also be important for signalling. This has been shown in the studies involving the IRAK-deficient cells. For the NF-κB signal to be restored, forms of IRAK which can become phosphorylated must be used (Li et al. 2001). Interestingly this is not the case for JNK activation, implying that IRAK phosphorylation is required only for the NF-κB signal.

Precisely how IRAK is recruited to the receptor complex is still not fully understood, although as stated above it probably involves death domain interactions. A protein termed Tollip (Toll interacting protein) has been described which can interact with IRAK (Burns et al. 2000), and may be involved in delivering IRAK to IL-1RAcP, which has been shown to interact in a yeast two-hybrid assay with IRAK (Volpe et al. 1997). Tollip has also been implicated in signalling by TLR-2 and TLR-6 (Bulut et al. 2001). Following IRAK recruitment, another adapter protein, called TNF receptor-associated factor (TRAF)6 comes into play; this is a member of the TRAF family of proteins implicated in TNF and CD40 signalling (Cao et al. 1996b) Recently it has been shown that TRAF6 occurs in a complex with two ubiquitinating enzymes Ubc13 and Uev1A (Wang et al. 2001). Once activated, TRAF6 becomes ubiquitinated, and interacts with three proteins termed TAB-1, TAB-2 and the MAP kinase TAK-1. IRAK has recently been shown to be required for the translocation of TAB-2 from the membrane to the TAK-1/TAB-1 complex (Takaesu et al. 2001). TAK-1 then becomes ubiquitinated (Wang et al. 2001). TAK-1, TAB-1 and TAB-2 also all become phosphorylated (QIAN et al. 2001). TAK-1 therefore undergoes both ubiquitination and phosphorylation. This unexpected round of ubiquitination is of some interest, as it does not lead to the degradation of the ubuquitinated proteins. Instead it appears to be a covalent modification akin to phosphorylation which activates TAK-1, thereby starting kinase cascades. One of these involves NF-κB, as TAK-1 can activate IKK-2, the kinase responsible for I-κB phosphorylation. I-κBα is the main form of I-κB and becomes phosphorylated on Ser32 and Ser36. This leads to its ubiquitination, which this time does lead to degradation. NF-κB is then released (largely the heterodimer p50/p65) and translocates to the nucleus, where it binds consensus sequences on target genes and up-regulates their expression.

A second pathway has also been shown to be required for NF-κB activation. This involves the low molecular weight G protein Rac, which becomes activated in response to IL-1 (JEFFERIES and O'NEILL 2000). Rac regulates a pathway culminating in enhanced transactivating activity of p65, and occurs downstream of MyD88 and IRAK (JEFFERIES et al. 2001). An identical phenomenon has been shown to occur in TLR-2 signalling (ARBIBE et al. 2000). Phosphatidylinositol 3-kinase (PI 3-kinase) has also been implicated in this pathway, and evidence for recruitment of the p85 subunit of PI 3-kinase to both IL-1RAcP and TLR-2 has been reported (ARBIBE et al. 2000; SIZEMORE et al. 1999).

The other kinase cascades activated by ubiquitinated TAK-1 involve MAP kinases, with TAK-1 activating upstream kinases for p38 and JNK (WANG et al. 2001). Both of these kinases phosphorylate transcription factors such as ATF-2, but possibly more importantly, are required for the stabilization of mRNAs which contain AU repeats (RIDLEY et al. 1997; CHEN et al. 1998). These include mRNAs for TNF and COX-2. A possible substrate for p38 here is a protein termed tristetraproline, which binds AU repeats and is active when phosphorylated (MAHTANI et al. 2001). This protein becomes phosphorylated by p38 during IL-1 signalling, and then binds to and stabilizes the untranslated mRNA for TNF.

We and others have found an unexpected role for the low molecular weight G protein Ras in IL-1 signalling (PALSSON et al. 2000; CAUNT et al. 2001). We have shown that Ras activation is required for the activation of p38 MAP kinase but not of NF-κB. Ras occurs downstream of TRAF6. It is possible that Ras and Rac are involved in IL-1 signalling because of the focal adhesion complex (FAC). IL-1RI has been shown to be recruited into FACs, as has IRAK (HUANG et al. 2001; MACGILLIVRAY et al. 2000). FACs form at points of contact between integrins and the extracellular matrix. Both Ras and Rac and indeed many other signalling proteins are found in FACs, and their participation in IL-1 signalling may be as a consequence of the IL-1RI complex localizing to FACs.

As can therefore be seen, a highly complex array of protein–protein interactions are involved in IL-1 signal transduction. Many of these are also likely to be involved in signal transduction by TLRs.

5 TLR Signal Transduction: Lessons from Knock-Out Mice and the Discovery of Mal

NF-κB, p38 and JNK have all been implicated in the signalling pathways for TLR-2, TLR-4 and TLR-9, largely through measuring their activation or the use of inhibitors. In addition, the use of dominant negative forms of MyD88, IRAK and TRAF6 have suggested a role for these proteins in signalling by these TLRs and also TLR-5. Another approach to implicate signalling proteins in TLR action is to examine the effect of TLR ligands in various knock-out mice. Mice deficient in MyD88, IRAK and TRAF6 have been described (Adachi et al. 1998; Kawai et al. 1999; Thomas et al. 1999; Naito et al. 1999; Kaisho et al. 2001). Their responses to IL-1, IL-18 and various TLR agonists are impaired, confirming a role for these signalling proteins in TIR domain-dependent signalling. Important differences emerged in these studies which led us to the discovery of Mal, a specific signal for TLR-4. Signals tested in response to IL-1, IL-18, TLR-2 or TLR-9 agonists, be they in vitro or in vivo, are severely impaired, in MyD88 knock-out mice. The results obtained with LPS however were more subtle. LPS was not lethal, probably because it was unable to induce TNF production. However, when macrophages were analysed for NF-κB, p38 and JNK, it was observed that the cells could still respond, although the response was somewhat delayed (Kawai et al. 1999). In addition, the maturation of dendritic cells induced by LPS was normal, as was induction of several interferon-sensitive genes (Kaisho et al. 2001; Kawai et al. 2001). This was not the case for CpG DNA (the ligand for TLR-9) which was totally impaired in these responses (Kaisho et al. 2001). These results implied that an additional adapter was present. Two groups working independently reported on a protein which may be this adapter (Fitzgerald et al. 2001; Horng et al. 2001). As mentioned above it is a homologue of MyD88, and is termed MyD88 adapter-like (Mal) (Fitzgerald et al. 2001) or TIRAP (Horng et al. 2001). Mal was found during a search of an expressed sequence tag library prepared from dendritic cells. It has a TIR domain but intriguingly lacks a death domain, having instead an unassigned domain in its N terminus. Mal over-expression mimics MyD88, activating NF-κB and JNK. It can form homodimers or heterodimers with MyD88. Importantly, unlike MyD88, Mal does not required IRAK, but does interact with IRAK-2. Its effects also require TRAF6 and TAK-1. Mal has been clearly implicated in LPS and TLR-4 signalling. Dominant-negative forms inhibit NF-κB activation by LPS or TLR-4. This is not the case when IL-1, IL-18 or TLR-2 are used, consistent with Mal being responsible for the signals observed in MyD88-deficient macrophages. In the study on TIRAP (Horng et al. 2001), similar results were obtained in terms of TLR-4, and no role was found in TLR-9 signalling. Importantly, TIRAP was shown to be involved in the activation of RNA-dependent protein kinase, the kinase responsible for responses to viral double-stranded RNA. Finally, a peptide based on the BB loop from TIRAP inhibited dendritic cell maturation, further pointing to its role as the adapter in TLR-4 signalling responsible for effects on these cells. Figure 3 compares the signalling pathway for TLR-4 and TLR-2. As can be seen, LPS action also requires CD14 and MD2, the latter being indispensable for LPS action.

Whether Mal plays a role in signalling by other TLRs is not known. A contender is TLR-3, which has been shown to recognize double-stranded RNA from viruses. TLR-3 signalling is not impaired in MyD88-deficient mice or in dendritic cells cultured from them (ALEXOPOULOU et al. 2001). Mal may therefore be an adapter for TLR-3.

The potential importance of Mal/TIRAP can be seen from studies comparing TLR-2 and TLR-4 action. TLR-2 and TLR-4 agonists will induce over-lapping sets of genes, although often with different intensities (RE and STROMINGER 2001). In addition, the expression of specific genes has been reported, such as *iNOS* and *IP10* in response to LPS and *IL-12p40* in response to peptidoglycan (RE and STROMINGER 2001; JONES et al. 2001; HIRSCHFIELD et al. 2001) as shown in Fig. 3. IP10 is an example of a cytokine important for Th1 cells whereas IL-12p40 inhibits T helper (Th)1 responses and promotes Th2 cell function. This implies that the adaptive immune response activated will be driven by specific TLRs, with signalling pathways governing the response. Mal is the only known difference in terms of signal transduction between TLR-2 and TLR-4 and may therefore be responsible

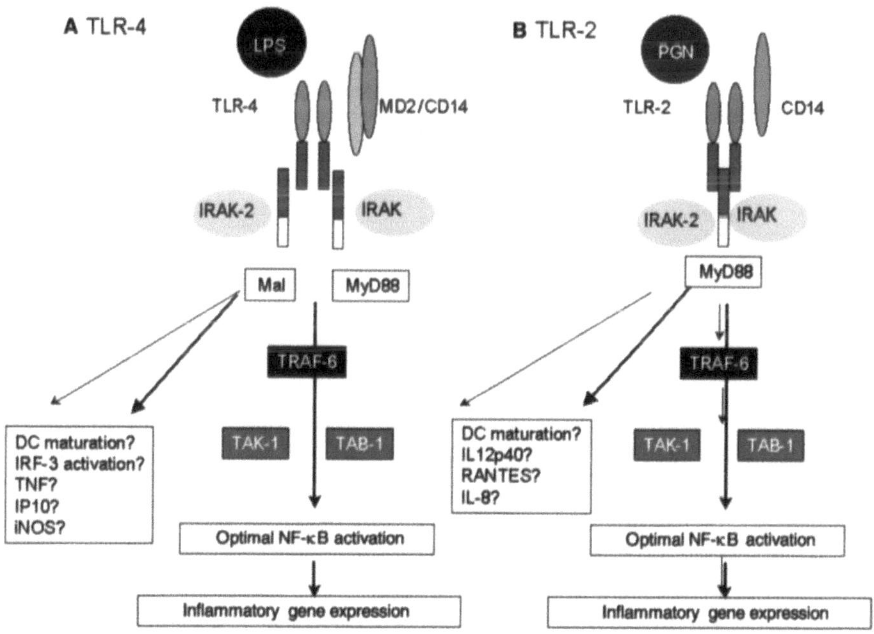

Fig. 3. Comparing TLR-4 and TLR-2 signal transduction. MyD88, IRAK, IRAK-2, TRAF6, TAB-1 and TAK-1 are involved in signalling by both TLR-4 and TLR-2, leading to activation of NF-κB. In the case of TLR-4, Mal is also required and may be the basis for differences in responses to TLR-4 and TLR-2 ligands, which lead to induction of specific genes as indicated. Mal may also be involved in the induction of dendritic cell maturation by TLR-4. Both Mal and MyD88 are required for optimal NF-κB activation by TLR-4. See text for details

for the different outcomes in terms of the genes induced. It may emerge that similar to the fly, different pathogens will be sensed by different TLRs, triggering different signalling pathways which culminate in pathogen-directed gene products being induced, leading to pathogen elimination. More global analyses of gene expression changes induced by pathogens or their products would concur with this model (Huang et al. 2001).

6 Conclusions

As the intricacies of signal transduction pathways for TLRs emerge, innate immunity will no longer be considered a non-specific response. The clear role TLRs play in pathogen recognition, and from there, the governing of adaptive responses represents an important breakthrough in the study of host defence, heralding in a new and exciting era for the study of innate immunity. The importance of signal transduction pathways in these processes is clear, as the specificities that are being uncovered are due not only to pathogen recognition but also to TLR-specific signalling cascades being activated. Further studies will provide additional clarity.

Notes added in proof. Recently, it has been shown that the anti-viral compound R-848 may be a ligand for TLR-7 and TLR-8 (Hemmi et al. (2002) Nature Immunol 3, 196–200; Jurk et al. (2002) Nature Immunol 3, 499). These compounds may therefore mimic viral PAMPs. Secondly, a novel kinase has been described, IRAK-4, which most interestingly may be the most receptor-proximal kinase activated by IL-1R and TLRs (Suzuki et al. (2002) Nature 416, 750–6; Li et al. (2002) PNAS USA 99, 5567–5572). IRAK-4 was shown to phosphorylate and activate IRAK, initiating the signalling cascade.

References

Adachi O, Kawai T, Takeda K, Matsumoto M, Tsutsui H, Sakagami M, Nakanishi K, Akira S (1998) Targetted disruption of the MyD88 gene results in loss of IL-1 and IL-18-mediated function. Immunity 9:143–150

Alexopoulou L, Holt AC, Medzhitov R, Flavell RA (2001) Recognition of double stranded RNA and activation of NF-κB by Toll-like receptor 3. Nature 413:696–712

Anderson KV, Bokla L, Nusslein-Volhard C (1985) Establishment of dorsal-ventral polarity in the *Drosophila* embryo: the induction of polarity by the *Toll* gene product. Cell 42:791–798

Aliprantis, AO, Yang RB, Mark MR, Suggett S, Devaux B, Radolf JD, Klimpel GR, Godowski P, Zychlinsky A (1999) Cell activation and apoptosis by bacterial lipoproteins through Toll-like receptor-2. Science 285:736–739

Arbibe L, Mira JP, Teusch N, Kline L, Guha M, Mackman N, Godowski PJ, Ulevitch RJ, Knaus UG (2000) Toll-like receptor 2 –mediated NF-κB requires a Rac1-dependent pathway. Nature Immunol 1:533–540

Bowie, A, Kiss-Toth E, Symons JA, Smith GL, Dower SK, O'Neill LA (2000) A46R and A52R from vaccinia virus are antagonists of host IL-1 and toll-like receptor signaling. Proc Natl Acad Sci USA 97:10162–10167

Brightbill HD, Libraty DH, Krutzik SR, Yang RB, Belisle JT, Bleharski JR, Maitland M, Norgard MV, Plevy SE, Smale ST, Brennan PJ, Bloom BR, Godowski PJ, Modlin RL (1999) Host defense mechanisms triggered by microbial lipoproteins through Toll-like receptors. Science 285:732–736

Bulut, Y, Faure E, Thomas L, Equils O, Arditi M (2001) Coperation of TLR-2 and -6 for cellular activation by soluble tuberculosis factor and *Borellia burdoferi* outer surface protein A lipoprotein: role of Toll-interacting protein and IL-1 receptor signalling molecules in Toll-like receptor 2 signalling. J Immunol 167:987–994

Burns, K, Martinon F, Esslinger C, Pahl H, Schneider P, Bodmer JL, Di Marco F, French L, Tschopp J (1998) MyD88: an adapter protein involved in interleukin-1 signalling. J Biol Chem 273:12203–12209

Burns, K, Clatworthy J, Martin L, Martinon F, Plumpton C, Maschera B, Lewis A, Ray K, Tschopp J, Volpe F (2000) Tollip, a new component of the IL-1RI pathway links IRAK to the IL-1 receptor. Nature Cell Biol 2:364–351

Cao Z, Henzel WJ, Gao X (1996a) IRAK: a kinase associated with the interleukin-1 receptor. Science 271:1128–1131

Cao Z., Xiong J, Takeuchi M, Kurama T, Goeddel DV (1996b) Traf-6 is a signal transducer for interleukin-1. Nature 383:443–446

Caunt CJ, Kiss-Toth E, Carlotti F, Chapman R, Qwarnstrom EE (2001) Ras controls tumor necrosis factor receptor-associated factor (TRAF)6-dependent induction of nuclear factor-κ b. Selective regulation through receptor signaling components. J Biol Chem 276:6280–6288

Chen CY, Del Gatto-Konczak F, Wu Z, Karin M (1998) Stabilization of interleukin-2 mRNA by the c-Jun NH2-terminal kinase pathway. Science 280:1945–1949

Dunn E, Sims JE, Nicklin MJ, O'Neill LA (2001) Annotating genes with potential roles in the immune system: six new members of the IL-1 family. Trends in Immunol 22:533–536

Da Silva Correia J, Soldau K, Christen U, Tobias PS, Ulevitch RJ (2001) Lipopolysaccharide is in close proximity to each of the proteins in its membrane receptor complex. Transfer from CD14 to TLR-4 and ND2. J Biol Chem 276:21129–21135

Fitzgerald KA, Palsson-McDermott EM, Bowie AG, Jefferies CA, Mansell AS, Brady G, Brint E, Dunne A, Gray P, Harte MT, McMurray D, Smith DE, Sims JE, Bird TA, O'Neill LA (2001) Mal (MyD88 adapter-like) is required for TLR-4 signal transduction. Nature 413:78–83

Gay NJ, Keith F (1991) Drosophila Toll and IL-1 receptor. Nature 351:355–356

Gewirtz AT, Navas TA, Lyons S, Godowski PJ, Madara JL (2001) Cutting edge: bacterial flagellin activates basolaterally expressed TLR-5 to induce epithelial proinflammatory gene expression. J Immunol 167:1882–1885

Hayashi F, Smith KD, Ozinsky A, Hawn TR, Yi EC, Goodlett DR, Eng JK, Akira S, Underhill DM, Aderem A (2001) The innate immune response to bacterial flagellin is mediated by Toll-like receptor 5. Nature 410:1099–1103

Heguy A, Baldari CT, Macchia G, Telford JL, Melli M (1992) Amino acids conserved in the interleukin-1 receptors (IL-1Rs) and the Drosophila Toll protein are essential for IL-1R signal transduction. J Biol Chem 267:2605–2609

Hedengren M, Asling B, Dushay MS, Ando I, Ekengren S, Wihlborg M, Hultmark D (1999) Relish, a central factor in the control of humoral but not cellular immunity in Drosophila. Mol Cell 4:1–20

Hemmi H, Takeuchi O, Kawai T, Kaisho T, Sato S, Sanjo H, Matsumoto M, Hoshino K, Wagner H, Takeda K, Akira S (2000) A Toll-liker receptor recognises bacterial DNA. Nature 408:740–745

Hirschfeld M, Weis JJ, Toshchakov V, Salkowski CA, Cody MJ, Ward DC, Qureshi N, Michalek SM, Vogel SN (2001) Signaling by toll-like receptor 2 and 4 agonists results in differential gene expression in murine macrophages. Infect Immun 69:1477–82

Horng T, Barton GM, Medzhitov R (2001) TIRAP: an adapter molecule in Toll signalling. Nature Immunol 2:835–841

Hoshino K, Takeuchi O, Kawai T, Sanjo H, Ogawa T, Takeda Y, Takeda K, Akira S (1999) Cutting edge: Toll-like receptor 4 (TLR4)-deficient mice are hyporesponsive to lipopolysaccharide: evidence for TLR4 as the *Lps* gene product. J Immunol 162:3749–3752

Huang Q, Liu D, Majewski, P, Schulte LC, Korn JM, Young RA, Lander ES and Hacohen N (2001) The plasticity of dendritic cell responses to pathogens and their components. Science 294:870–875

Huang WY, Valles S, Qwarnstrom EE (2001) Translocation of the IL-1 receptor to focal adhesions is regulated through the C-terminal end of the cytoplasmic domain. Cell Biol Int 25:309–317

Janssens, S et al (2002) Regulation of interleukin-1-induced NF-κB activation by alternative splicing of MyD88. Current Biol (in press)

Jefferies C, O'Neill LAJ (2000) Rac1 regulated IL-1-induced NF-κB in an I-κB-independent manner by enhancing the ability of the p65 subunit to transactivate gene expression. J Biol Chem 275:3114–3120

Jefferies C, Bowie A, Brady G, Cooke EL, Li X, O'Neill LA (2001) Transactivation by the p65 subunit of NF-κB in response to IL-1 involves MyD88, IRAK, Traf-6 and Rac1. Mol Cell Biol 21:4544–4552

Jones BW, Means TK, Heldwein KA, Keen MA, Hill PJ, Belisle JT, Fenton MJ (2001) Different Toll-like receptor agonists induce distinct macrophage responses. J Leukoc Biol 69:1036–1044

Kaisho T, Takeuchi O, Kawai T, Hoshino K, Akira S (2001) Endotoxin-induced maturation of MyD88-deficient dendritic cells. J Immunol 166:5688–5694

Kawai, T, Adachi O, Ogawa T, Takeda K, Akira S (1999) Unresponsiveness of MyD88-deficient mice to endotoxin. Immunity 11:115–150

Kawai T, Takeuchi O, Fujita T, Inoue J, Muhlradt PF, Sato S, Hoshino K, Akira S (2001) Lipopolysaccharide stimulates the MyD88-independent pathway and results in activation of IFN-regulatory factor 3 and the expression of a subset of lipopolysaccharide-inducible genes. J Immunol 167:5887–5894

Kidd S (1992) Characterisation of the *Drosophila cactus* locus and analysis of interaction between cactus and dorsal proteins. Cell 71:623–635

Li X, Commane M, Burns C, Vithalani K, Cao Z, Stark G (1999) Mutant cells that do not respond to IL-1 reveal a novel role for IL-1 receptor-associated kinase. Mol Cell Biol 19:4643–4652

Lemaitre B, Nicolas E, Michaut L, Reichhart JM, Hoffmann JA (1996) The dorsoventral regulatory gene cassette spatzle/Toll/cactus controls the potent anti-fungal response in *Drosophila* adults. Cell 86:973–983

Levashina EA, Langley E, Green C, Gubb D, Ashburner M, Hoffmann JA, Reichhart JM (1999) Constitutive activation of toll-mediated antifungal defense in serpin-deficient Drosophila. Science 285:1917–1919

Letsou A, Alexander S, Orth K, Wasserman SA (1991) Genetic and molecular characterisation of *tube*, a *Drosophila* gene maternally required for embryonic dorsoventral polatiry. Proc Natl Acad Sci USA 88:810–814

Lord KA, Hoffman-Liebermann B, Liebermann DA (1990) Nucleotide sequence and expression of a cDNA encoding MyD88, a novel myeloid differentiation primary response gene induced by IL6.Oncogene 5:1095–1097

MacGillivray MK, Cruz TF, McCulloch CA (2000) The recruitment of the interleukin-1 (IL-1) receptor-associated kinase (IRAK) into focal adhesion complexes is required for IL-1beta -induced ERK activation. J Biol Chem 275:23509–23515

Mahtani KR, Brook M, Dean JL, Sully G, Saklatvala J, Clark AR (2001) Mitogen-activated protein kinase p38 controls the expression and posttranslational modification of tristetraprolin, a regulator of tumor necrosis factor alpha mRNA stability. Mol Cell Biol 21:6461–6469

Mansell A, Reinicke A, Worrall DM, O'Neill LAJ (2002) The serine protease inhibitor anti-thrombin-III inhibits LPS-mediated NF-κB activation by TLR-4. FEBS Lett 508:313–317

Medzhitov R, Preston-Hurlburt P, Janeway CA Jr (1997) A human homologue of the *Drosophila* Toll protein signals activation of adaptive immunity. Nature 388:394–397

Medzhitov R, Preston-Hurlburt P, Kopp E, Stadlen A, Chen C, Ghosh S, Janeway CA Jr (1998) MyD88 is an adapter in the hToll/Il-1 receptor family signalling pathways. Mol Cell 2:253–258

Muzio M, Li J, Feng P, Dixit V (1997) IRAK (Pelle) family member IRAK-2 and MyD88 as proximal mediators of IL-1 signaling. Science 278:1612–1615

Naito A, Azuma S, Tanaka S, Miyazaki T, Takaki S, Takatsu K, Nakao K, Nakamura K, Katsuki M, Yamamoto T, Inoue J (1999) Severe osteopetrosis, defective interleukin-1 signalling and lymph node organogenesis in TRAF6-deficient mice. Genes Cells 1999 4:353–362

O'Neill LAJ (2000) the interleukin-1 receptor / Toll-like receptor superfamily: signal transduction during inflammation and host defense. Science's Signal Transduction Knowledge Environment. http://www.stke.org/cgi/content/full/OC_sigtrans;2000/44/re1

O'Neill LAJ, Greene C (1998) Signal transduction pathways activated by the IL-1 receptor family: Ancient signalling machinery in mammals, insects and plants. J Leuk Biol 63:650–657

Ozinsky A, Underhill DM, Fontenot JD, Hajjar AM, Smith KD, Wilson CB, Schroeder L, Aderem A (2000) The reportoire for pattern recognition of pathogens by the innate immune system is defined by co-operation between Toll-like receptors. Proc Natl Acad Sci USA 97:13766–13771

Palsson EM, Popoff M, Thelestam M, O'Neill LA (2000) Divergent roles for Ras and Rap in the activation of p38 mitogen-activated protein kinase by interleukin-1. J Biol Chem 275:7818–7825

Poltorak A, He X, Smirnova I, Liu MY, Huffel CV, Du X, Birdwell D, Alejos E, Silva M, Galanos C, Freudenberg M, Ricciardi-Castagnoli P, Layton B, Beutler B (1998) Defective LPS signalling in C3H/HeJ and C57BL/10ScCr mice: mutations in *Tlr4* gene. Science 282:2085–2088

Pujol N, Link EM, Liu LX, Kurz CL, Alloing G, Tan M, Ray KP, Solari R, Johnson CD, Ewbank JJ (2001) A reverse genetic analysis of components of the Toll signaling pathway in Caenorhabditis elegans. Curr Biol 11:809–821

Qian Y, Commane M, Ninomiya-Tsuji J, Matsumoto K, Li X (2001) IRAK-mediated translocation of Traf-6 and TAB2 in the interleukin-1-induced activation of NF-κB. J Biol Chem 276:41661–41667

Qureshi ST, Lariviere L, Leveque G, Clermont S, Moore KJ, Gros P, Malo D (1999) Endotoxin-tolerent mice have mutations in Toll-like receptor 4 (*Tlr4*). J Exp Med 189:615–625

Re F, Strominger JL (2001) Toll-like receptor 2 (tlr2) and tlr4 differentially activate human dendritic cells. J Biol Chem 276:37692–37699

Ridley SH, Sarsfield SJ, Lee JC, Bigg HF, Cawston TE, Taylor DJ, DeWitt DL, Saklatvala J (1997) Actions of IL-1 are selectively controlled by p38 mitogen-activated protein kinase: regulation of prostaglandin H synthase-2, metalloproteinases, and IL-6 at different levels. J Immunol 158: 3165–3173

Roth S, Stein D, Nusslein-Volhard C (1989) A gradient of nuclear localisation of the dorsal protein determines dorsoventral patterns in the Drosophila embryo. Cell 59:1189–1202

Schreuder H, Tardif C, Trump-Kallmeyer S, Soffientini A, Sarubbi E, Akeson A, Bowlin T, Yanofsky S, Barrett RW (1997) A new cytokine-receptor binding mode revealed by the crystal structure of the IL-1 receptor with an antagonist. Nature 386:194–200

Schwandner R, Dziarski R, Wesche H, Rothe M, Kirschning CJ (1999) Peptidoglycan- and lipoteichoic acid-induced cell activation is mediated by Toll-like receptor-2. J Biol Chem 274:17406–17409

Shelton CA, Wasserman, SA (1993) *pelle* encodes a protein kinase required to establish dorsoventral polarity in the *Drosophila* embryo. Cell 72:515–525

Sims JE, March CJ, Cosman D, Widmer MB, MacDonald HR, McMahan CJ, Grubin CE, Wignall JM, Jackson JL, Call SM, Dower SK (1988) cDNA expression cloning of the IL-1 receptor, a member of the immunoglobulin superfamily. Science 241:585–589

Sims JE, Dower SK (1994) Interleukin-1 receptors. Eur Cyt Netw 5:539–546

Sizemore N, Leung S, Stark GR (1999) Activation of PI 3-kinase in response to IL-1 leads to phosphorylation and activation of the NF-kappa p65/RelA subunit. Mol Cell Biol 19:4798–4805

Takaesu, G., Ninomiya-Tsuji J, Kishida S, Li X, Stark GR, Matsumoto K (2001) Interleukin-1 (IL-1) receptor-associated kinase leads to activation of TAK-1 by inducing TAB2 translocation in the IL-1 signalling pathway. Mol Cell Biol 21:2475–2484

Takeuchi, O, Hoshino K, Kawai T, Sanjo H, Takada H, Ogawa T, Takeda K, Akira S (1999) Differential roles of TLR2 and TLR4 in recognition of Gram-negative and Gram-positive bacterial cell wall components. Immunity 11:443–451

Takeuchi, O Kaufmann A, Grote K, Kawai T, Hoshino K, Morr M, Muhlradt PF, Akira S (2000) Cutting edge: preferentially the R-stereoisomer of the mycoplasmal lipopeptide macrophage-activating lipopeptide-2 activates immune cells through a toll-like receptor-2- and MyD88-dependent signalling pathway. J Immunol 164:554–557

Takeuchi O, Kawai T, Muhlradt PF, Morr M, Radolf JD, Zychlinsky A, Takeda K, Akira S (2001) Discrimination of bacterial lipoproteins by Toll-like receptor 6. Int Immunol 13:933–940

Thomas JA, Allen JL, Tsen M, Dubnicoff T, Danao J, Liao XC, Cao Z, Wasserman SA (1999) Impaired cytokine signaling in mice lacking the IL-1 receptor-associated kinase. J Immunol 163:978–984

Underhill, D, Ozinsky A, Hajjar AM, Stevens A, Wilson CB, Bassetti M, Aderem A (1999) Toll-like receptor-2 is recruited to macrophage phagosomes and discriminates between pathogens. Nature 401:1477–1482

Volpe, F, Clatworthy J, Kaptein A, Maschera B, Griffin AM, Ray K (1997) The IL-1 receptor accessory protein is responsible for the recruitment of the interluakin-1 receptor associated kinase to the IL-1/IL-1 receptor complex. FEBS Lett 419:41–44

Wang C, Deng L, Hong M, Akkaraju GR, Inoue J, Chen ZJ (2001) TAK1 is a ubiquitin-dependent kinase of MKK and IKK. Nature 412:346–351

Wesche H, Henzel WJ, Shillinglaw W, Li S, Cao Z (1997) MyD88: an adapter that recruits IRAK to the IL-1 receptor complex. Immunity 7:837–847

Wu LP, Anderson KV (1998) regulated nuclear import of Rel proteins in the *Drosophila* immune response. Nature 395:93–97

Wesche H, Gao X, Li X, Kirschning CJ, Stark GR, Cao Z (1999) IRAK-M is a novel member of the Pelle/interleukin-1 receptor-associated kinase (IRAK) family. J Biol Chem 274:19403–19410

Xu Y, Tao X, Shen B, Horng T, Medzhitov R, Manley JL, Tong L (2000) Structural basis for signal transduction by the Toll/interleukin-1 receptor domains. Nature 408:111–115

Yamin TT and Miller DK (1997) The Interleukin-1 receptor-associated kinase is degraded by proteasomes following its phosphorylation. J Biol Chem 272:21540–21547

Yang RB, Mark MR, Gurney AL, Godowski PJ (1999) Signaling events induced by lipopolysaccharide-activated toll-like receptor. J Immunol 163:639–643

Toll Receptors in *Drosophila*: a Family of Molecules Regulating Development and Immunity

J.L. Imler and J.A. Hoffmann

In recent years, Toll-like receptors (TLRs) have emerged as key receptors which detect microbes and initiate an inflammatory response. The Toll receptor was originally identified and characterized 14 years ago for its role in the embryonic development of the fruit-fly *Drosophila melanogaster*. Subsequently, it was also shown to be an essential component of the signaling pathway mediating the anti-fungal host defense in this model organism. New factors involved in the activation of the Toll receptor or in intracytoplasmic signaling during the immune response in *Drosophila* have recently been identified. The existence of significant functional differences between mammalian TLRs and *Drosophila* Toll receptors is also becoming apparent.

1 Discovery of the Toll Receptor and of Its Importance for the Immune Response

The *Toll* mutant was discovered in the early 1980s by C. Nüsslein-Volhard and K. Anderson, during a mutagenesis screen aimed at unravelling the genes involved in the establishment of the dorso-ventral (DV) axis of the *Drosophila* embryo (Anderson and Nüsslein-Volhard 1984). This screen identified a total of 12 genes that are required for the determination of the DV axis of the embryo. The *Toll* gene was cloned in 1988 and found to encode a new type of transmembrane receptor (Hashimoto et al. 1988). Its extracellular domain is composed of leucine-rich repeats flanked by characteristic cystein-rich motifs, whereas its intracyto-

UPR9022 du CNRS, Institut de Biologie Moléculaire et Cellulaire, 15 rue Descartes, 67000 Strasbourg, France, *e-mail: JL.Imler@ibmc.u-strasbg.fr*

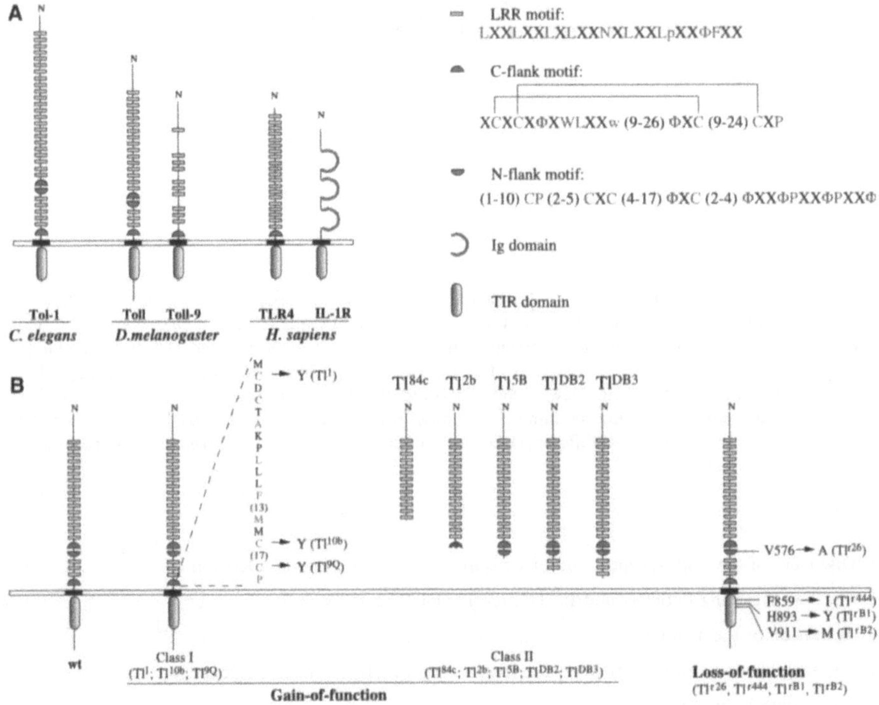

Fig. 1A,B. Structure of TIR domain receptors. **A** Comparison of TIR domain receptors in *C. elegans*, *Drosophila* and humans. Φ, Hydrophobic residue. The position of the disulfide bridges in the C-flank motif were determined in GpIb alpha. **B** Structure of the products of gain-of-function (ventralizing) and loss-of-function (dorsalizing) alleles of *Toll* isolated by saturating mutagenesis of the third chromosome in *Drosophila*. Class I gain-of-function alleles are dominant, whereas class II alleles require a wild-type copy of Toll to induce ventralization

plasmic domain did not resemble any structure known at the time (Fig. 1). Independent contemporary cloning of the interleukin-1 receptor (IL-1R) revealed an undescribed intracytoplasmic domain, connected to an ectodomain with three immunoglobulin domains (Sims et al. 1988). This domain appeared to show striking similarities with the cytoplasmic tail of Toll, and was thereafter referred to as the TIR (Toll/IL-IR) homology domain (Gay and Keith 1991). Of the 11 other genes of the pathway, seven (*pipe*, *windbeutel*, *nudel*, *gastrulation defective*, *snake*, *easter* and *spätzle*) code for factors required for activation of Toll, and four (*tube*, *pelle*, *cactus* and *dorsal*) code for factors acting downstream of the receptor (reviewed in Anderson 2000; LeMosy et al. 1999) (Figs. 2, 3). Loss-of-function mutations in these genes result in dorsalization of the embryo, with the exception of mutations in the *cactus* gene, which lead to ventralization of the embryos. Interestingly, both dorsalizing and ventralizing alleles of Toll were recovered in this screen (hence the name Toll, which is a German slang term for 'fantastic') (Anderson 2000). Three of the ventralizing alleles, which correspond to gain-of-function mutations yielding a constitutively active receptor, carry cysteine to

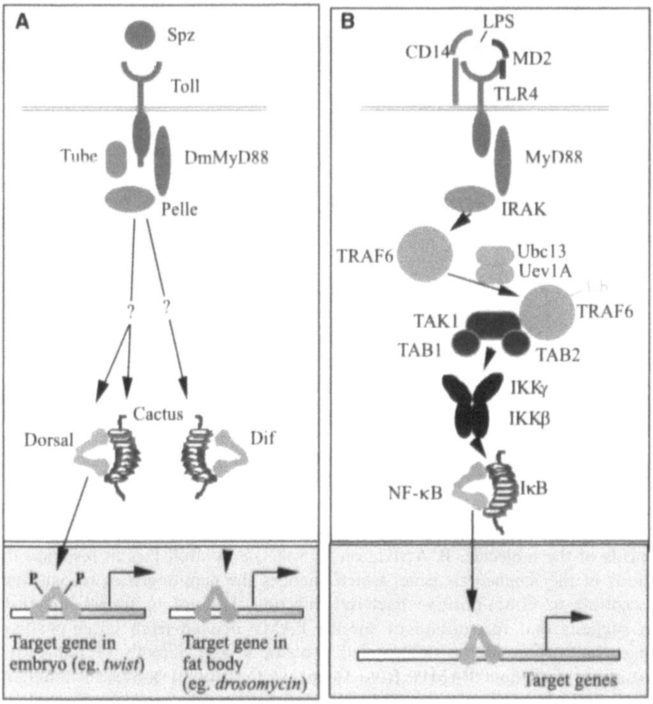

Fig. 2A,B. The Toll/TLR signaling pathways. **A** In *Drosophila*, Toll is activated by the cysteine-knot growth factor Späetzle. The receptor complex contains the adapter proteins Tube and DmMyD88, as well as the kinase Pelle. Activation of Toll triggers phosphorylation and degradation of the inhibitor Cactus by the proteasome, thus releasing the Rel proteins Dorsal and Dif. Phosphorylation of Dorsal on conserved serine residues is required for nuclear translocation. **B** In mammals, TLR4 associates with the coreceptors CD14 and MD2 to bind LPS derived from Gram-negative bacteria. The kinase IRAK is recruited to the receptor complex through its association with the adapter protein MyD88. IRAK activates TRAF6 which becomes ubiquitinated by the factors Ubc13 and Uev1A. Ubiquitinated TRAF6 activates the TAK1/TAB1/TAB2 complex, and the TAK1 kinase activates the IKK complex. IKKβ phosphorylates IκB and triggers its degradation by the proteasome (reviewed in Silverman and Maniatis 2001)

tyrosine mutations in a membrane-proximal cysteine-rich motif of the ectodomain. The other ventralizing alleles encode truncated soluble versions of the receptor, and induce ventralization only in the presence of a wild-type copy of *Toll* (Schneider et al. 1991) (Fig. 1). This suggests that the truncated protein associates with the normal receptor and that this association of the ectodomains is sufficient to trigger signaling. Activation of the Toll pathway on the ventral side of the embryo results in a ventral-to-dorsal gradient of nuclear translocation of the Dorsal transcription factor. High, intermediate and low levels of nuclear Dorsal lead to differential regulation of a selected set of genes, thus establishing embryonic polarity (Belvin and Anderson 1996 and references therein).

The demonstration of an immune function of the Toll receptor came from a set of completely unrelated studies. It had long been known that the host defense of some insects involves secretion into the hemolymph of a battery of potent

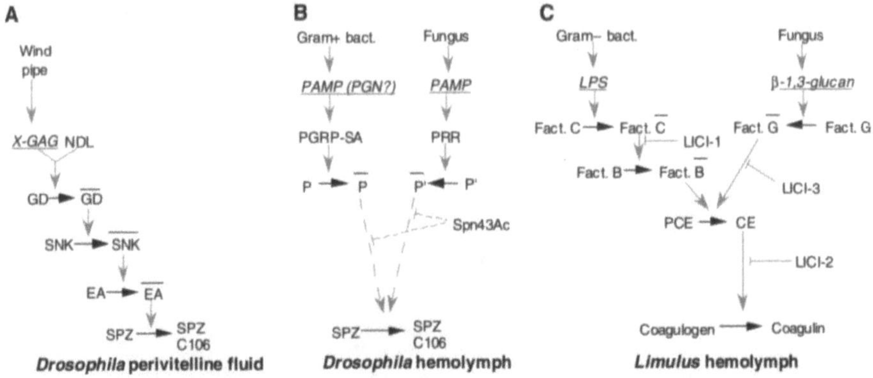

Fig. 3A–C. Proteolytic cascades activating Späetzle in *Drosophila* and the related molecule Coagulogen in *Limulus*. **A** Activation of Späetzle in the perivitelline fluid on the ventral side of the *Drosophila* embryo. The heparan-sulfate 2-O sulfotransferase Pipe modifies a glycosaminoglycan (*GAG*) chain on an unidentified factor (*X*), which is thought to act in concert with Nudel (*NDL*) to trigger autocatalytic activation of Gastrulation defective (*GD*). This serine protease then activates Snake (*SNK*), which cleaves Easter (*EA*). Easter then cleaves Späetzle (*SPZ*) and releases a cysteine-knot growth factor composed of the 106 C-terminal amino acids of the molecule. **B** Activation of Späetzle in adult flies in response to infection. Flies with mutations in the *semmelweis* gene, which encodes the peptidoglycan recognition protein PGRP-SA, are susceptible to Gram-positive bacterial infection, but not to fungal infection (Michel et al. 2001). This suggests that recognition of specific PAMP derived from Gram-positive bacteria by PGRP-SA triggers activation of the unidentified protease P and, directly or indirectly, cleavage of Späetzle. Recognition of distinct PAMPs from fungi are thought to activate a different protease (P'). In flies bearing mutations in the *nec* gene, which encodes the serpin Spn43Ac, Späetzle is constitutively present in its active, C106, form. **C** Activation of blood coagulation in the horseshoe crab. The zymogens factor C and factor G are activated by different PAMPs derived from Gram-negative bacteria and fungi, respectively. Conversion of proclotting enzyme (*PCE*) into clotting enzyme (*CE*), results in cleavage of Coagulogen and formation of Coagulin, which is structurally related to Späetzle C106. Coagulin molecules assemble to form an insoluble gel which prevents spreading of infectious microorganisms (Muta and Iwanaga 1996). The serpins LICI-1, LICI-2 and LICI-3 control these proteolytic cascades. Non-self or modified-self patterns are represented in *underlined italics*, and serpins are indicated with *shadowed characters*

cationic anti-microbial peptides (reviewed in Hoffmann and Reichhart 1997). Some of these molecules, namely Drosomycin and Metchnikowin, are active against various fungi, but lack activity against other microbes. Others are active against Gram-negative (Drosocin, Diptericin and Cecropin) or Gram-positive (Defensin) bacteria. These peptides are synthesized in response to infection by a functional equivalent of the mammalian liver, the fat body. The study of the regulation of the expression of these genes provided a powerful system to address the signaling pathways activating this immune response. Classical promoter mapping experiments with the *cecropin A1* and *diptericin* genes in cell lines and in transgenic flies quickly revealed the importance of DNA motifs closely related to mammalian nuclear factor (NF)-κB binding sites (Engstrom et al. 1993; Meister et al. 1994). At the time when these data became available, the only member of the Rel family of transcription factors (to which NF-κB belongs) identified in Drosophila was encoded by the gene *dorsal* mentioned above. When *dorsal* mutant flies were immune-challenged, induction of anti-microbial peptide gene expression was found to be wild-type, suggesting that another Rel protein mediates induction of the anti-microbial peptide genes (see below) (LEMAITRE et al. 1995b). Analysis with other mutants of the Toll pathway showed that the *spätzle-Toll-tube-pelle-cactus* gene cassette is required to control *drosomycin* gene expression (LEMAITRE et al. 1996). In particular, flies carrying mutations in these genes are highly susceptible to infection by fungi and Gram-positive bacteria, although their resistance to infection by Gram-negative bacteria is not affected (LEMAITRE et al. 1996; RUTSCHMANN et al. 2002). None of the other genes that define the embryonic Toll pathway was subsequently found to be required for the control of anti-microbial peptide gene expression in adult flies. Following these discoveries 5 years ago, major advances have been made in the characterization of the Toll-like receptor (TLR) family in mammals (see other chapters in this volume). In *Drosophila*, attention has focused on the activation of the Toll receptor in response to infection, as described below. A family of Toll molecules has also been described, and these molecules show some structural and functional differences from mammalian TLRs. We will further discuss the strong indications that the response to Gram-negative bacterial infections is largely independent of Toll and differs markedly from that of mammals.

2 The Toll Signaling Pathway and Its Transcriptional Targets

As observed for many other receptors devoid of catalytic activity, the intracytoplasmic domain of the Toll receptor associates with a complex of signaling molecules. Genetic screens first identified the *tube* and *pelle* genes as being required for the specification of the DV axis (ANDERSON and NÜSSLEIN-VOLHARD 1984). *Tube* encodes a 50-kDa protein composed of an amino-terminal death domain and a C-terminal region characterized by the presence of five repeats of an eight-residue motif (N-Φ-P-X-Φ-T/S-X-L where Φ is a hydrophobic, and X any, residue) (LETSOU et al. 1991). This feature is not shared by any other protein predicted to be encoded by the *Drosophila* genome, although single copies of this motif can be identified in several *Drosophila* proteins with diverse functions. *Pelle* encodes a homologue of the IL-1R associated kinase (IRAK), and is composed of an amino-terminal death domain and a C-terminal serine–threonine kinase domain (SHELTON and WASSERMAN 1993). Both proteins interact through their death domains (SCHIFFMANN et al. 1999; XIAO et al. 1999). Elegant genetic experiments using injection of mRNA encoding gain-of-function versions of *tube* and *pelle* in *Drosophila* embryos have shown that Tube acts upstream of Pelle (GALINDO et al. 1995; GROSSHANS et al. 1994). However, the mechanism by which the signal is transduced from the Toll receptor through Tube and Pelle to the nucleus remains largely unknown. The role of Tube in particular is mysterious, as it does not appear to interact directly with the intracytoplasmic domain of Toll (SHEN and MANLEY

1998). By contrast, Pelle binds directly through its kinase domain to a region within the intracytoplasmic domain of Toll. Thus, Tube does not appear to function as a simple adapter in the Toll receptor complex. Importantly, residue V899 (valine in position 899) of the Toll intracytoplasmic domain, which corresponds to the critical proline residue P674 in TLR4 (Poltorak et al. 1998), is not required for binding Pelle, but is necessary for Toll function (Xu et al. 2000). This suggests that other molecules interact with the intracytoplasmic domain of Toll. One of these molecules has recently been identified as a *Drosophila* homologue of MyD88 (DmMyD88) (Horng and Medzhitov 2001; Tauszig-Delamasure et al. 2002). DmMyD88 can be coimmunoprecipitated in transfected S2 cells with Toll, Tube and Pelle, indicating that it is a component of the receptor complex. In addition, over-expression of DmMyD88 in cell lines and transgenic flies results in constitutive expression of *drosomycin*. Furthermore, expression of dominant-negative versions of this molecule interferes with the Toll-mediated activation of *drosomycin*. Finally, induction of *drosomycin* expression is strongly reduced in *DmMyD88* mutant flies, and these flies rapidly succumb to fungal and Gram-positive bacterial infections (Tauszig-Delamasure et al. 2002). Epistasy experiments place DmMyD88 upstream of Tube and Pelle in the Toll pathway. Thus, the *Drosophila* Toll receptor complex differs from the mammalian TLR complexes by the presence of an additional component, that is, Tube. The importance of this molecule may be related to the presence of an additional domain in the intracytoplasmic tail of Toll (Fig. 1A). This C-terminal extension, present in several other members of the *Drosophila* Toll family (Tauszig et al. 2000), has been proposed to function as an inhibitory domain (Shen and Manley 1998). Although this has not been proven experimentally, activation of Toll is thought to activate the Pelle kinase. Importantly, the catalytic activity of Pelle is required for its biological activity, as kinase-dead versions of the enzyme do not rescue the dorsalized phenotype in mutant embryos (Towb et al. 2001). In vitro experiments have shown that Pelle can autophosphorylate. Pelle can also phosphorylate Tube and the intracytoplasmic domain of Toll, although the functional significance of these modifications is not completely understood (Grosshans et al. 1994; Shen and Manley 1998). Recent data indicate that the Tube phosphorylation causes down-regulation of the signal-dependent aggregation of Tube at the plasma membrane, and may therefore represent a feedback mechanism (Towb et al. 2001). We have a better understanding of the significance of Pelle autophosphorylation, as phosphorylated Pelle cannot associate with Toll (Shen and Manley 1998). This modification of Pelle may therefore allow the kinase to dissociate from the receptor complex, a situation reminiscent of the phosphorylation-induced release of IRAK from TLR4/IL-1R in mammals (Wesche et al. 1997). The receptor complex may also include the cytoskeleton protein Filamin (Edwards et al. 1997), and the evolutionary conserved factor Pellino (Grosshans et al. 1999), which have been identified in yeast two-hybrid screens as potential binding partners of Toll and Pelle, respectively. In the absence of fly lines mutant for these genes, the significance of these results remains unclear, although it is interesting to note that Pellino does not interact with catalytically inactive versions of Pelle.

Activation of the Toll receptor complex, either on the ventral side of the embryo, or in larvae and adults upon immune challenge, leads to phosphorylation and degradation of the IκB homologue Cactus, which retains the Rel proteins Dorsal and Dif in the cytoplasm of unstimulated cells (BELVIN and ANDERSON 1996; NICOLAS et al. 1998) (Fig. 2). The link between Pelle and Cactus phosphorylation still remains to be established. In mammals, the phosphorylated form of the Pelle homologue kinase IRAK interacts with tumor necrosis factor (TNF) receptor-associated factor (TRAF)6 (Fig. 2) (CAO et al. 1996). Three genes encoding TRAF factors have been identified in the *Drosophila* genome. The first, dTRAF1, exhibits strong sequence similarity with TRAF4 (which cannot activate NF-κB), and has been shown to interact with the Ste20 kinase Misshapen and to activate the Jun kinase JNK in mammalian cells (LIU et al. 1999). Independent experiments using mammalian cell culture assays have also shown that dTRAF1 is able to interact with members of the TNFR family such as p75 NTR and LTβR. Moreover, dTRAF1 and Pelle can be co-immunoprecipitated when they are over-expressed in mammalian cells, and their coexpression results in induction of NF-κB (ZAPATA et al. 2000). The second *Drosophila* TRAF molecule, dTRAF2, shows strongest sequence similarity with mammalian TRAF6. It interacts with Pelle and can be phosphorylated by this kinase in vitro. In addition, dTRAF2 has been shown to synergize with Pelle to activate the transcription factor Dorsal in transfected *Drosophila* S2 cells (SHEN et al. 2001). dTRAF2 also interacts with dECSIT (Evolutionary Conserved Signaling Intermediate in Toll pathways), a factor which activates expression of *diptericin* and *defensin* when over-expressed in transfected *Drosophila* cells (KOPP et al. 1999). In mammalian cells, ECSIT has been proposed to bridge TRAF6 and the kinase MEKK-1. Similar types of experiments in vitro or using transfected cells tentatively identified the kinase DLAK, which is structurally related to IKKβ, as the Cactus kinase (KIM et al. 2000). However, the subsequent isolation of fly strains mutant for the gene encoding this kinase (*ird5*) revealed that it is not required for inactivation of the inhibitor Cactus (see below) (LU et al. 2001). Thus, most of the in vitro experiments described above should be interpreted with caution, and the roles of the *Drosophila* TRAF molecules in the Toll pathway remain conjectural at this stage.

The signal-induced degradation of Cactus allows the release and nuclear translocation of the two related Rel factors Dif (*Drosophila* immunity factor) and Dorsal in fat-body cells (IP et al. 1993; LEMAITRE et al. 1995b). In the *Drosophila* embryo, phosphorylation of a specific conserved serine residue of Dorsal is also required for establishment of the DV axis. This Toll-mediated post-translational modification of Dorsal controls the nuclear translocation of the transcription factor independently of Cactus (DRIER et al. 2000) (Fig. 2). It is not yet known if a similar mechanism operates to control Dif nuclear translocation. As mentioned above, the immune-inducibility of the genes encoding anti-microbial peptides is not affected in *dorsal* mutant flies. By contrast, induction of *drosomycin* expression is abolished in *Dif* mutant flies (MANFRUELLI et al. 1999; MENG et al. 1999; RUTSCHMANN et al. 2000a, 2002). Interestingly, in larvae, Dorsal can substitute for Dif and mediate up-regulation of *drosomycin* expression suggesting the existence of different cofactors operating with the Rel proteins during the immune response in larvae and adults (PETERSEN et al. 1999). Because *dorsal* is expressed in the adult fat-body, and is strongly up-regulated by septic injury (LEMAITRE et al. 1995b), it may be involved in the regulation of the other transcriptional targets of the Toll pathway such as the *Drosophila* immune induced molecules that have been identified in the hemolymph of challenged flies (UTTENWEILER-JOSEPH et al. 1998).

3 Activation of the Toll Pathway

The *spätzle* gene functions immediately upstream of *Toll* in the DV pathway in the embryo and it encodes a member of the large family of cystein-knot growth factors (McDONALD and HENDRICKSON 1993). Späetzle is synthesized as a 41-kDa precursor which forms an 82-kDa covalently linked homodimer (DELOTTO and DELOTTO 1998; MORISATO and ANDERSON 1994). This factor is proteolytically processed to its active form, composed of the carboxy-terminal 106 amino acids. This activated 24-kDa dimer is thought to bind to the Toll receptor, thus triggering its dimerization and activation. A serine protease cascade culminates in the activation of Späetzle in the embryo and involves the zymogens encoded by the genes *nudel*, *gastrulation defective*, *snake*, and *easter* (Fig. 3A) (LEMOSY et al. 1999). There is clear biochemical evidence that Gastrulation defective activates Snake, which in turn activates Easter (DISSING et al. 2001; LEMOSY et al. 2001). This latter protease cleaves Späetzle (DELOTTO and DELOTTO 1998). Regarding activation of Gastrulation defective itself, the current data available do not indicate that Nudel activates Gastrulation defective, but rather suggest that Gastrulation defective is activated by autocleavage. These data favor a model in which the Nudel protease activates a cofactor, which in turn activates Gastrulation defective (DISSING et al. 2001; HAN et al. 2000). The last two genes in the DV pathway, *windbeutel* and *pipe*, encode molecules of the secretory pathway. The *pipe* gene encodes a heparan-sulfate 2-O sulfotransferase which is probably involved in the modification of proteoglycans in the Golgi, whereas *windbeutel* encodes an homologue of the endoplasmic reticulum protein ERp29, which is required for the correct subcellular distribution of Pipe (SEN et al. 2000). Thus, Pipe might be involved in the modification of the cofactor recognized and cleaved by Nudel. Consistent with this hypothesis, Nudel is a secreted glycoprotein with a composite structure reminiscent of extracellular matrix proteins (HONG and HASHIMOTO 1995), and Gastrulation defective and Snake have been shown to interact with heparin–sepharose (DISSING et al. 2001).

During the immune response, activation of the Toll pathway depends on Späetzle, but not on the serine protease cascade operating during embryonic development. However, activation of the pathway involves at least one serine protease, as unprocessed Späetzle cannot activate Toll-dependent *drosomycin* expression (J.L. Imler and J.A. Hoffmann, unpublished results), and cleavage of

Späetzle is observed in immune-challenged flies. Moreover, mutant flies bearing mutations in the *necrotic* gene, which encodes the serine protease inhibitor (serpin) Spn43AC, constitutively express *drosomycin* (LEVASHINA et al. 1999). Thus, it is likely that recognition of infection by fungi or Gram-positive bacteria activates proteolytic cascades, which culminate in the cleavage and activation of Späetzle (Fig. 3). This process is conceptually reminiscent of the activation of blood coagulation in another invertebrate, the horseshoe crab. Elegant biochemical experiments have shown that in this system, recognition of lipopolysaccharide (LPS) or β-glycans by the zymogens factor C and factor G, respectively, leads to the activation of two cascades which converge on another zymogen, proclotting enzyme (MUTA and IWANAGA 1996) (Fig. 3C). The resulting clotting enzyme converts soluble coagulogen to an insoluble coagulin gel, which prevents spreading of the infecting microorganisms. Interestingly, the crystal structure of coagulogen revealed that the C-terminal half of the molecule adopts a typical cystein-knot growth factor fold, like Späetzle and neurotrophins (BERGNER et al. 1996). In addition, factor B and proclotting enzyme share with Snake and Easter a small compact domain with three disulfide bonds, known as the clip domain, situated in the *amino*-terminal parts of the molecules, in addition to a carboxy-terminal serine protease catalytic domain (MUTA and IWANAGA 1996).

This conceptual similarity has recently received experimental support. Indeed, a genetic screen aimed at identifying genes involved in the control of *drosomycin* gene expression has identified a mutant strain, *semmelweis* (*seml*), in which induction of *drosomycin* is strongly reduced upon infection by Gram-positive bacteria, but not by fungi. In addition, these flies were shown to be highly susceptible to infection by Gram-positive bacteria, whereas they have the same sensitivity to fungal infection as wild-type flies (MICHEL et al. 2001). These results demonstrate the existence of two distinct pathways operating upstream of Späetzle in the immune response (Fig. 3B). Interestingly, the *seml* gene encodes a known peptidoglycan recognition protein (PGRP-SA) (WERNER et al. 2000). Injection of wild-type hemolymph into *seml* flies is sufficient to rescue the activation of the Toll pathway upon infection with the Gram-positive bacteria *Micrococcus luteus*. This demonstrates that PGRP-SA is a circulating protein in the hemolymph, where it can detect Gram-positive bacterial infection (MICHEL et al. 2001). Thus, whereas in mammals microbial components can be directly recognized by TLRs, during the systemic response in *Drosophila* recognition of infectious non-self occurs in the blood, and triggers a proteolytic cascade which generates an endogenous Toll ligand (Fig. 3).

4 The Response to Gram-Negative Bacterial Infection Is Largely Toll-Independent

As early as in 1995, it became apparent that the regulation of the antimicrobial response in *Drosophila* involves at least two distinct pathways. Indeed, characterization of the first immunosuppressed *Drosophila* mutant, *immune deficiency* (*imd*), revealed that the immune inducibility of the genes encoding antibacterial peptides was abolished or strongly reduced in *imd* flies (LEMAITRE et al. 1995a). In contrast, the inducible expression of the anti-fungal peptide Drosomycin was not affected in this mutant background. This result was confirmed 1 year later when it was shown that the inducible synthesis of the antibacterial peptides Drosocin and Diptericin is essentially normal in *Toll* mutant flies (LEMAITRE et al. 1996). Genetic screens conducted in several independent laboratories have recently identified four other genes encoding components of the Imd pathway (Fig. 4) (ELROD-ERICKSON et al. 2000; LEULIER et al. 2000; LU et al. 2001; RUTSCHMANN et al. 2000b; VIDAL et al. 2001). This pathway activates the third member of the NF-κB family in *Drosophila*, Relish (HEDENGREN et al. 1999). Unlike Dorsal and Dif, Relish is synthesized as an inactive precursor, which is maintained in the cytoplasm by an ankyrin repeat-containing inhibitory domain. This is reminiscent of the mammalian factors NF-κB1 (p100) and NF-κB2 (p105). However, unlike these factors, activation of Relish does not involve the proteasome (STÖVEN et al. 2000). Rather, Relish is activated by endoproteolytic cleavage. This cleavage may be mediated by a caspase which remains to be identified, as the linker region of Relish, which connects the Rel domain to the ankyrin repeats, contains a potential caspase cleavage site (SILVERMAN and MANIATIS 2001). The fact that ectopic

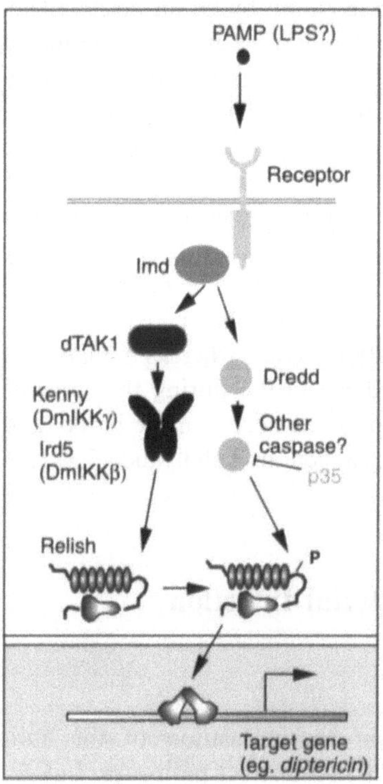

Fig. 4. The Imd pathway which controls the response to Gram-negative infection in *Drosophila*. The p105-like transcription factor Relish is activated upon phosphorylation by the *Drosophila* IKK complex, which contains the kinase DmIKKβ, encoded by the *ird5* gene, and the regulatory subunit DmIKKγ, encoded by *kenny*. Activation of Relish involves an endoprotease, which may be a caspase, to release the Rel domain from the inhibitory ankyrin repeats. Ectopic expression of the viral caspase inhibitor p35 blocks induction of the genes encoding antibacterial peptides. Dredd is a caspase-8 homologue which is not sensitive to inhibition by p35. dTAK1 is a homologue of the TAK1 kinase, and Imd contains a death domain which is most closely related to the death domain of the adapter RIP, which associates with the TNF receptor

expression of the viral caspase inhibitor p35 abolishes the immune inducibility of the antibacterial peptide genes also points to the importance of caspases in this pathway (GEORGEL et al. 2001). However, the caspase Dredd, which is required in the Imd pathway, does not appear to cleave Relish directly (ELROD-ERICKSON et al. 2000; LEULIER et al. 2000; SILVERMAN and MANIATIS 2001). Dredd, which is closely related to Caspase-8, an apical caspase, and is not sensitive to p35, possibly activates an effector caspase to cleave Relish. Activation of Relish involves phosphorylation of this factor by the product of the *ird5* gene, which encodes a homologue of IKKβ (SILVERMAN et al. 2000). Similar to its mammalian homologue, DmIKKβ is associated with a regulatory subunit (DmIKKγ), encoded by the *kenny* gene, which is required for its function (RUTSCHMANN et al. 2000b; SILVERMAN et al. 2000). Upstream of this signaling complex, the homologue of TAK1 (transforming growth factor-β activated kinase) is also required for induction of the antibacterial peptides (VIDAL et al. 2001). Finally, the *imd* gene encodes the most upstream component of the pathway identified to date (GEORGEL et al. 2001). *Imd* encodes an apparent adapter protein with a death domain which resembles that of RIP, a mammalian factor that bridges the TNF receptor to the IκB kinase (IKK) complex (SILVERMAN and MANIATIS 2001). Thus, the Imd pathway appears to be composed of homologues of molecules involved in the IL-1/TLR pathway (TAK1), the TNF pathway (Caspase-8, RIP), or both (IKKβ, IKKγ, NF-κB).

The receptor activating the Imd pathway remains unknown. It is an open question whether this receptor functions in a manner similar to Toll, i.e., in response to an endogenous ligand activated by a proteolytic cascade in the hemolymph, or is directly activated by recognition of microbial ligands. In support of the latter possibility, induction of anti-microbial peptide gene expression in surface epithelia by bacterial infection is dependent on the Imd pathway (ONFELT TINGVALL et al. 2001; TZOU et al. 2000). The identification of TLR4 as the long sought LPS signaling receptor (POLTORAK et al. 1998) suggested at one point that the receptor triggering the Imd pathway in response to Gram-negative bacterial infection was a member of the Toll family. Indeed, the *Drosophila* genome was found to encode eight additional receptors related to Toll: 18-Wheeler (18W) and Toll-3 to Toll-9 (TAUSZIG et al. 2000). With the exception of Toll-9, these receptors differ from mammalian TLRs by the presence of additional C-flank cystein rich motifs, and the presence of the distinct N-flank motif (Fig. 1). In addition, several members of the family (Toll, 18W, Toll-6, -7 and -8) contain a C-terminal intracytoplasmic extension following the TIR domain. Sequence comparison of the TIR domains confirms the existence of two subgroups of Toll receptors, the Toll subgroup containing *Drosophila* Toll, 18W, Toll-3 to Toll-8 and *Caenorhabditis elegans* Tol-1; and the TLR subgroup, containing the ten mammalian TLRs described to date, as well as *Drosophila* Toll-9 (DU et al. 2000; LUO and ZHENG 2000). Apart from Toll, there are no indications so far that the other members of the *Drosophila* family participate in the control of the immune response. In 1997, loss-of-function mutant *18w* larvae were reported to exhibit severely reduced inducibility of the gene encoding the antibacterial peptide Attacin in response to Gram-negative bacterial infection (WILLIAMS et al. 1997). This initial study was based on the $18w^{7-35}$ mutation, which generates a truncated protein lacking the extracellular and the transmembrane domains (Eldon et al. 1994). *18w*$^{7-35}$ mutants die at the end of the last larval instar, prior to metamorphosis. Recent data show that, when the *18w*$^{7-35}$ mutation is combined with a chromosomal deletion (Df) which uncovers the *18w* gene, viable adults can be obtained, indicating that the lethality of *18w*$^{7-35}$ larvae does not result from the lack of a functional *18w* product. Analysis of the immune response in these flies (*18w*$^{7-35}$/Df) shows that induction of anti-microbial peptides in response to Gram-negative bacterial infection is similar to wild-type (P. Ligoxygakis and J.M. Reichhart, personal communication). Therefore, these data are not in favor of a selective role for the 18-Wheeler receptor in sensing Gram-negative infection to mount an appropriate response. In addition, over-expression of constitutively active forms of *Drosophila* Toll receptors either alone or in combination, is not sufficient to mimic the effects of LPS treatment in tissue culture cells (Imler et al. 2000; Tauszig et al. 2000). Moreover, inhibition of these receptors by double-stranded RNA interference (Clemens et al. 2000) does not affect the capacity of cells to respond to LPS or bacteria (S. Tauszig-Delamasure and J. L. Imler, unpublished results). The recent identification of a mutant strain for the adapter DmMyD88 also confirmed that recognition of Gram-negative bacterial infection proceeds differently in *Drosophila* and in mammals. Indeed, whereas *myd88* mutant mice do not respond to any of the ligands activating the TLRs, including LPS, *DmMyD88* mutant flies respond normally to Gram-negative bacterial infection (Takeuchi et al. 2000; Tauszig-Delamasure et al. 2002). Thus, our current data do not favor the hypothesis that in *Drosophila* Toll receptors operate in the control of Gram-negative bacterial infections. Rather, most *Drosophila* Tolls may serve developmental functions, as already described for Toll and 18W (Eldon et al. 1994; Halfon and Keshishian 1998; Qiu et al. 1998; Rose et al. 1997). This hypothesis is supported by the fact that several Toll receptors are strongly expressed during embryogenesis and metamorphosis (Tauszig et al. 2000). These receptors may be activated by other members of the *Drosophila* Späetzle gene family (Späetzle-2, to -6) (Parker et al. 2001). Alternatively, some members of the Toll family may function as adhesion molecules (Eldon et al. 1994; Keith and Gay 1990).

5 Concluding Remarks

Toll receptors were initially discovered in *Drosophila*, where their role in innate immunity was first described. These molecules are able to trigger a rapid transcriptional response by acting on preformed transcription factors of the Rel family, which are retained in the cytosol by inhibitory factors. This rapid response is essential to enable host cells to control rapidly proliferating infectious microorganisms. It is also a powerful means of providing rapid positional information to cells during development, which may explain why this system also became co-opted during *Drosophila* embryonic development. Despite the existence of striking similarities between the molecular control of innate immunity in flies and humans, the data we have reviewed here point to the existence of important differences between Toll receptors in these groups. First, phylogenetic analysis reveals the existence of two subfamilies of Toll receptors, the Tolls and the TLRs. Whereas most or all TLRs in mammals appear to be involved in innate immunity, this does not appear to be the case for Tolls. Indeed, in *Drosophila* 18 W, Toll-3, -4, -6, -7, and -8 may activate targets distinct from Rel proteins. This is evocative of the situation described for the Tol-1 receptor in *C. elegans*, an organism which does not contain any Rel protein coding gene (Pujol et al. 2001). Second, in *Drosophila* Toll does not function like TLRs do in mammals. Whereas TLRs appear to function as pattern recognition receptors and are activated upon association with microbial ligands, activation of Toll is mediated by an endogenous ligand. Hence, Toll is functionally more closely related to the other type of TIR domain receptors in mammals, that is the IL-1Rs, than to TLRs. As discussed above, it is now established that the systemic immune response in *Drosophila* is activated in the hemolymph by soluble pattern recognition receptors, which may reflect physiological differences between mammals and insects. Proteolytic cascades can amplify the signal generated upon local recognition of infectious non-self in the hemolymph which bathes all *Drosophila* organs, and may be necessary to efficiently trigger synthesis and systemic release of anti-microbial peptides by the fat-body. Third, the response to Gram-negative bacterial infection in *Drosophila* and mammals appears to use different mechanisms, and does not seem to involve Toll receptors. Rather, the Imd pathway involves factors resembling molecules operating in the TNF-α signaling pathway in mammals, such as Imd (RIP), Dredd (Caspase-8), Ird5 (IKKβ), Kenny (IKKγ) and Relish (NK-κB). The reason for the existence of a distinct pathway to control Gram-negative bacterial infection in *Drosophila* is not clear at this stage. The identification and functional characterization of the receptor activating this pathway, and of its putative mammalian homologue, represents a major challenge for the coming months.

Acknowledgements. We thank C. Hetru and J.-M. Reichhart for discussions, and H. Bilak and J. Royet for critical reading of the manuscript.

References

Anderson KV (2000) Toll signaling pathways in the innate immune response. Curr Opin Immunol 12:13–19

Anderson KV, Nusslein-Volhard C (1984) Information for the dorsal-ventral pattern of the *Drosophila* embryo is stored as maternal mRNA. Nature 311:223–227

Belvin MP, Anderson KV (1996) A conserved signaling pathway: the *Drosophila* toll-dorsal pathway. Ann Rev Cell Dev Biol 12:393–416

Bergner A, Oganessyan V, Muta T, Iwanaga S, Typke D, Huber R, Bode W (1996) Crystal structure of a coagulogen, the clotting protein from horseshoe crab: a structural homologue of nerve growth factor. EMBO J 15:6789–6797

Cao Z, Xiong J, Takeuchi M, Kurama T, Goeddel DV (1996) TRAF6 is a signal transducer for interleukin-1. Nature 383:443–446

Clemens JC, Worby CA, Simonson-Leff N, Muda M, Maehama T, Hemmings BA, Dixon JE (2000) Use of double-stranded RNA interference in *Drosophila* cell lines to dissect signal transduction pathways. Proc Natl Acad Sci USA 97:6499–6503

DeLotto Y, DeLotto R (1998) Proteolytic processing of the *Drosophila Spatzle* protein by easter generates a dimeric NGF-like molecule with ventralising activity. Mech Dev 72:141–148

Dissing M, Giordano H, DeLotto R (2001) Autoproteolysis and feedback in a protease cascade directing *Drosophila* dorsal-ventral cell fate. EMBO J 20:2387–2393

Drier EA, Govind S, Steward R (2000) Cactus-independent regulation of Dorsal nuclear import by the ventral signal. Curr Biol 10:23–26

Du X, Poltorak A, Wei Y, Beutler B (2000) Three novel mammalian toll-like receptors: gene structure, expression, and evolution. Eur Cytokine Netw 11:362–371

Edwards DN, Towb P, Wasserman SA (1997) An activity-dependent network of interactions links the Rel protein Dorsal with its cytoplasmic regulators. Development 124:3855–3864

Eldon E, Kooyer S, D'Evelyn D, Duman M, Lawinger P, Botas J, Bellen H (1994) The *Drosophila* 18 wheeler is required for morphogenesis and has striking similarities to Toll. Development 120:885–899

Elrod-Erickson M, Mishra S, Schneider D (2000) Interactions between the cellular and humoral immune responses in *Drosophila*. Curr Biol 10:781–784

Engstrom Y, Kadayalil L, Sun S, Samakovlis C, Hultmark D, Faye I (1993) kB-like motifs regulate the induction of immune genes in *Drosophila*. J Mol Biol 232:327–333

Galindo RL, Edwards DN, Gillespie SK, Wasserman SA (1995) Interaction of the pelle kinase with the membrane-associated protein tube is required for transduction of the dorsoventral signal in *Drosophila* embryos. Development 121:2209–2218

Gay N, Keith F (1991) *Drosophila* Toll and IL-1 receptor. Nature 351:355–356

Georgel P, Naitza S, Kappler C, Ferrandon D, Zachary D, Swimmer C, Kopczynski C, Duyk G, Reichhart JM, Hoffmann JA (2001) *Drosophila* immune deficiency (IMD) is a death domain protein that activates the antibacterial response and can promote apoptosis. Development Cell 1(4):503–514

Grosshans J, Bergmann A, Haffter P, Nüsslein-Volhard C (1994) Activation of the kinase Pelle by Tube in the dorsoventral signal transduction pathway of *Drosophila embryo*. Nature 372:563–566

Grosshans J, Schnorrer F, Nusslein-Volhard C (1999) Oligomerisation of Tube and Pelle leads to nuclear localisation of dorsal. Mech Dev 81:127–138

Halfon MS, Keshishian H (1998) The Toll pathway is required in the epidermis for muscle development in the *Drosophila* embryo. Dev Biol 199:164–174

Han JH, Lee SH, Tan YQ, LeMosy EK, Hashimoto C (2000) Gastrulation defective is a serine protease involved in activating the receptor toll to polarize the *Drosophila* embryo. Proc Natl Acad Sci USA 97:9093–9097

Hashimoto C, Hudson K, Anderson K (1988) The Toll gene of *Drosophila*, required for dorsal-ventral embryonic polarity, appears to encode a transmembrane protein. Cell 52:269–279

Hedengren M, Asling B, Dushay M, Ando I, Ekengren S, Wihlborg M, Hultmark D (1999) Relish, a Central Factor in the Control of Humoral but Not Cellular Immunity in *Drosophila*. Molecular Cell 4:1–20

Hoffmann J, Reichhart J (1997) *Drosophila* immunity. Trends in Cell Biol 7:309–316

Hong CC, Hashimoto C (1995) An unusual mosaic protein with a protease domain, encoded by the nudel gene, is involved in defining embryonic dorsoventral polarity in *Drosophila*. Cell 82:785–794

Horng T, Medzhitov R (2001) *Drosophila* MyD88 is an adapter in the Toll signaling pathway. Proc Natl Acad Sci USA 98:12654–12658

Imler JL, Tauszig S, Jouanguy E, Forestier C, Hoffmann JA (2000) LPS-induced immune response in *Drosophila*. J Endotoxin Res 6:459–462

Ip Y, Reach M, Enstrom Y, Kadalayil L, Cai H, Gonzalez-Crespo S, Tatei K, Levine M (1993) Dif, a dorsal-related gene that mediates an immune response in *Drosophila*. Cell 75:753–763

Keith J, Gay N (1990) The *Drosophila* membrane receptor Toll can function to promote cellular adhesion. EMBO J 9:4299–4306

Kim YS, Han SJ, Ryu JH, Choi KH, Hong YS, Chung YH, Perrot S, Raibaud A, Brey PT, Lee WJ (2000) Lipopolysaccharide-activated kinase, an essential component for the induction of the antimicrobial peptide genes in *Drosophila melanogaster* cells. J Biol Chem 275:2071–2079

Kopp E, Medzhitov R, Carothers J, Xiao C, Douglas I, Janeway CA, Ghosh S (1999) ECSIT is an evolutionarily conserved intermediate in the Toll/IL-1 signal transduction pathway. Genes Dev 13:2059–2071

Lemaitre B, Kromer-Metzger E, Michaut L, Nicolas E, Meister M, Georgel P, Reichhart J, Hoffmann J (1995a) A recessive mutation, immune deficiency (imd), defines two distinct control pathways in the *Drosophila* host defense. Proc Natl Acad Sci USA 92:9365–9469

Lemaitre B, Meister M, Govind S, Georgel P, Steward R, Reichhart JM, Hoffmann JA (1995b) Functional analysis and regulation of nuclear import of dorsal during the immune response in *Drosophila*. EMBO J 14:536–545

Lemaitre B, Nicolas E, Michaut L, Reichhart J, Hoffmann J (1996) The dorsoventral regulatory gene cassette spätzle/Toll/cactus controls the potent antifungal response in *Drosophila* adults. Cell 86: 973–983

LeMosy EK, Hong CC, Hashimoto C (1999) Signal transduction by a protease cascade. Trends Cell Biol 9:102–107

LeMosy EK, Tan YQ, Hashimoto C (2001) Activation of a protease cascade involved in patterning the *Drosophila* embryo. Proc Natl Acad Sci USA 98:5055–5060

Letsou A, Alexander S, Orth K, Wasserman SA (1991) Genetic and molecular characterization of tube, a *Drosophila* gene maternally required for embryonic dorsoventral polarity. Proc Natl Acad Sci USA 88:810–814

Leulier F, Rodriguez A, Khush RS, Abrams JM, Lemaitre B (2000) The *Drosophila* caspase Dredd is required to resist Gram-negative bacterial infection. EMBO reports 1:353–358

Levashina EA, Langley E, Green C, Gubb D, Ashburner M, Hoffmann JA, Reichhart JM (1999) Constitutive activation of toll-mediated antifungal defense in serpin- deficient *Drosophila*. Science 285:1917–1919

Liu H, Su YC, Becker E, Treisman J, Skolnik EY (1999) A *Drosophila* TNF-receptor-associated factor (TRAF) binds the ste20 kinase Misshapen and activates Jun kinase. Curr Biol 9:101–104

Lu Y, Wu LP, Anderson KV (2001) The antibacterial arm of the *Drosophila* innate immune response requires an IκB kinase. Genes Dev 15:104–110

Luo C, Zheng L (2000) Independent evolution of Toll and related genes in insects and mammals. Immunogenetics 51:92–98

Manfruelli P, Reichhart JM, Steward R, Hoffmann JA, and Lemaitre B (1999) A mosaic analysis in *Drosophila* fat body cells of the control of antimicrobial peptide genes by the Rel proteins and DIF. EMBO J 18:3380–3391

McDonald N, Hendrickson W (1993) A structural superfamily of growth factors containing a cystine knot motif. Cell 73:421–424

Meister M, Braun A, Kappler C, Reichhart JM, Hoffmann JA (1994) Insect immunity. A transgenic analysis in *Drosophila* defines several functional domains in the diptericin promoter. EMBO J 13:5958–5966

Meng X, Khanuja BS, Ip YT (1999) Toll receptor-mediated *Drosophila* immune response requires Dif, an NF- kappaB factor. Genes Dev 13:792–797

Michel T, Reichhart JM, Hoffmann JA, Royet J (2001) *Drosophila* Toll is activated by Gram-positive bacteria through a circulating peptidoglycan recognition protein. Nature 414:756–759

Morisato D, Anderson K (1994) The spätzle gene encodes a component of the extracellular signaling pathway establishing the dorsal-ventral pattern of the *Drosophila* embryo. Cell 76:677–688

Muta T, Iwanaga S (1996) The role of hemolymph coagulation in innate immunity. Curr Opin Immunol 8:41–47

Nicolas E, Reichhart J, Hoffmann J, Lemaitre B (1998) *In vivo* regulation of the IκB homologue cactus during the immune response of *Drosophila*. J Biol Chem 273:10463–10469

Onfelt Tingvall T, Roos E, Engstrom Y (2001) The imd gene is required for local Cecropin expression in *Drosophila* barrier epithelia. EMBO Rep 2:239–243

Parker JS, Mizuguchi K, Gay NJ (2001) A family of proteins related to Spatzle, the toll receptor ligand, are encoded in the *Drosophila* genome. Proteins 45:71–80

Petersen UM, Kadalayil L, Rehorn KP, Hoshizaki DK, Reuter R, Engstrom Y (1999) Serpent regulates *Drosophila* immunity genes in the larval fat body through an essential GATA motif. EMBO J 18:4013–4022

Poltorak A, He X, Smirnova I, Liu M, Huffel C, Du X, Birdwell D, Alejos E, Silva M, Galanos C, Freudenberg M, Ricciardi-Castagnoli P, Layton B, Beutler B (1998) Defective LPS signaling in C3H/HeJ and C57BL/10ScCr mice: mutations in Tlr4 gene. Science 282:2085–2088

Pujol N, Link EM, Liu LX, Kurz CL, Alloing G, Tan M, Ray KP, Solari R, Johnson CD, Ewbank JJ (2001) A reverse genetic analysis of components of the Toll signaling pathway in *Caenorhabditis elegans*. Curr Biol 11:809–821

Qiu P, Pan PC, Govind S (1998) A role for the *Drosophila* Toll/Cactus pathway in larval hematopoiesis. Development 125:1909–1920
Rose D, Zhu X, Kose H, Hoang B, Cho J, Chiba A (1997) Toll, a muscle cell surface molecule, locally inhibits synaptic initiation of the RP3 motoneuron growth cone in *Drosophila*. Development 124:1561–1571
Rutschmann S, Jung A, Hetru C, Reichhart J, Hoffmann J, Ferrandon D (2000a) The Rel protein DIF mediates the Toll-dependent antifungal response in *Drosophila*. Immunity 12:569–580
Rutschmann S, Jung A, Zhou R, Silverman N, Hoffmann JA, Ferrandon D (2000b) Role of *Drosophila* IKKγ in a Toll-independent antibacterial immune response. Nature Immunol 1:342–347
Rutschmann S, Kilinc A, Ferrandon D (2002) The Toll pathway is required for resistance to Gram-positive bacterial infections in *Drosophila*. J Immunol 168:1542–1546
Schiffmann DA, White JH, Cooper A, Nutley MA, Harding SE, Jumel K, Solari R, Ray KP, Gay NJ (1999) Formation and biochemical characterization of tube/pelle death domain complexes: critical regulators of postreceptor signaling by the *Drosophila* toll receptor. Biochemistry 38:11722–11733
Schneider D, Hudson K, Lin T, Anderson K (1991) Dominant and recessive mutations define functional domains of Toll, a transmembrane protein required for dorsal-ventral polarity in the *Drosophila* embryo. Genes & Dev 5:797–807
Sen J, Goltz JS, Konsolaki M, Schupbach T, Stein D (2000) Windbeutel is required for function and correct subcellular localization of the *Drosophila* patterning protein Pipe. Development 127: 5541–5550
Shelton CA, Wasserman SA (1993) pelle encodes a protein kinase required to establish dorsoventral polarity in the *Drosophila* embryo. Cell 72:515–525
Shen B, Liu H, Skolnik EY, Manley JL (2001) Physical and functional interactions between *Drosophila* TRAF2 and Pelle kinase contribute to Dorsal activation. Proc Natl Acad Sci USA 98:8596–8601
Shen B, Manley JL (1998) Phosphorylation modulates direct interactions between the Toll receptor, Pelle kinase and Tube. Development 125:4719–4728
Silverman N, Maniatis T (2001) NF-κB signaling pathways in mammalian and insect innate immunity. Genes Dev 15:2321–2342
Silverman N, Zhou R, Stoven S, Pandey N, Hultmark D, Maniatis T (2000) A *Drosophila* IκB kinase complex required for Relish cleavage and antibacterial immunity. Genes Dev 14:2461–2471
Sims JE, March CJ, Cosman D, Widmer MB, MacDonald HR, McMahan CJ, Grubin CE, Wignall JM, Jackson JL, Call SM, et al. (1988) cDNA expression cloning of the IL-1 receptor, a member of the immunoglobulin superfamily. Science 241:585–589
Stöven S, Ando I, Kadalayil L, Engström Y, Hultmark D (2000) Activation of the *Drosophila* NF-κB factor Relish by rapid endoproteolytic cleavage. EMBO reports 1:347–352
Takeuchi O, Takeda K, Hoshino K, Adachi O, Ogawa T, Akira S (2000) Cellular responses to bacterial cell wall components are mediated through MyD88-dependent signaling cascades [In Process Citation]. Int Immunol 12:113–117
Tauszig S, Jouanguy E, Hoffmann JA, Imler JL (2000) Toll-related receptors and the control of antimicrobial peptide expression in *Drosophila*. Proc Natl Acad Sci USA 97:10520–10525
Tauszig-Delamasure S, Bilak H, Capovilla M, Hoffmann JA, Imler JL (2002) *Drosophila* MyD88 is required for the response to fungal and Gram-positive bacterial infections. Nat Immunol 3:91–97
Towb P, Bergmann A, Wasserman SA (2001) The protein kinase Pelle mediates feedback regulation in the *Drosophila* Toll signaling pathway. Development 128:4729–4736
Tzou P, Ohresser S, Ferrandon D, Capovilla M, Reichhart JM, Lemaitre B, Hoffmann JA, Imler JL (2000) Tissue-specific inducible expression of antimicrobial peptide genes in *Drosophila* surface epithelia. Immunity 13:737–748
Uttenweiler-Joseph S, Moniatte M, Lagueux M, Van Dorsselaer A, Hoffmann JA, Bulet P (1998) Differential display of peptides induced during the immune response of *Drosophila*: a matrix-assisted laser desorption ionization time-of- flight mass spectrometry study. Proc Natl Acad Sci USA 95:11342–11347
Vidal S, Khush RS, Leulier F, Tzou P, Nakamura M, Lemaitre B (2001) Mutations in the *Drosophila* dTAK1 gene reveal a conserved function for MAPKKKs in the control of rel/NF-κB-dependent innate immune responses. Genes Dev 15:1900–1912
Werner T, Liu G, Kang D, Ekengren S, Steiner H, Hultmark D (2000) A family of peptidoglycan recognition proteins in the fruit fly *Drosophila* melanogaster. Proc Natl Acad Sci USA 97: 13772–13777
Wesche H, Henzel WJ, Shillinglaw W, Li S, Cao Z (1997) MyD88: an adapter that recruits IRAK to the IL-1 receptor complex. Immunity 7:837–847

Williams M, Rodriguez A, Kimbrell D, Eldon E (1997) The *18-wheeler* mutation reveals complex antibacterial gene regulation in *Drosophila* host defense. EMBO J 16:6120–6130
Xiao T, Towb P, Wasserman SA, Sprang SR (1999) Three-dimensional structure of a complex between the death domains of Pelle and Tube. Cell 99:545–555
Xu Y, Tao X, Shen B, Horng T, Medzhitov R, Manley JL, Tong L (2000) Structural basis for signal transduction by the Toll/interleukin-1 receptor domains. Nature 408:111–115
Zapata JM, Matsuzawa S, Godzik A, Leo E, Wasserman SA, and Reed JC (2000) The *Drosophila* tumor necrosis factor receptor-associated factor-1 (DTRAF1) interacts with Pelle and regulates NFκB activity. J Biol Chem 275:12102–12107

Toll-Like Receptors and Their Ligands

G.M. Barton and R. Medzhitov

The Toll-like receptors (TLRs) are key molecules involved in the recognition of pathogens by the innate immune system. This family of germ line-encoded receptors has evolved to recognize conserved features of microbes. Currently, 10 TLR family members have been identified in mammals. The number of ligands for these receptors continues to grow, and it seems clear that multiple ligands exist for each receptor. Whether engagement of different TLRs leads to differences in gene expression and thereby differences in the immune response remains to be seen. However, recent work has demonstrated that activation of TLRs is required for initiation of only certain adaptive immune responses.

1 Recognition by the Innate Immune System

The central challenge faced by all immune systems is the discrimination of foreign, non-self from self. The ability to recognize and eliminate infectious microbes is essential for the survival of any multicellular organism. This task is daunting, however, considering the overwhelming diversity of potentially infectious

Section of Immunobiology, Howard Hughes Medical Institute, Yale University School of Medicine, 310 Cedar St., BML 458, New Haven, CT 06520, USA, *e-mail: gregory.barton@yale.edu*

organisms that the immune system must be able to detect. For the adaptive immune systems of vertebrates this challenge is met with the enormous range of specificities of clonally expressed antigen receptors on T and B cells. These receptors are generated randomly and subsequently evaluated based on their specificities for self. Cells expressing self-reactive cells are eliminated such that only potentially non-self recognizing receptors remain. The cost of this breadth of specificities is that the adaptive immune system is slow to respond to infection. It takes days for the specific T and B cells to mount a full response, a period of time in which many microbes would kill the host if unchecked. Therefore, the initial response mounted by the cells of the innate immune system is essential for survival of the host. This innate immune recognition leads to effector responses that help to control infections as well as to induction of signals that control the subsequent adaptive immune response.

Activation of the innate immune system of vertebrates is based on a strategy of recognition that reveals its ancient evolutionary roots. The mechanism of recognition must be inherently different from that of the adaptive immune system because it is not based on randomly generated antigen receptors with essentially infinite specificities. Instead, the innate immune system has evolved to recognize a limited number of conserved microbial features. These targets have been termed pathogen-associated molecular patterns (PAMPs), although they are not present exclusively on pathogens. In fact, they are often conserved products of microbial metabolism common to all microbes of a particular class [e.g., lipopolysaccharide (LPS) on Gram-negative bacteria]. PAMPs are recognized by evolutionarily conserved, germ line-encoded pattern-recognition receptors (PRRs). Certain types of these receptors can be found in the serum while others are expressed on phagocytic antigen presenting cells (APCs), such as macrophages and dendritic cells (DCs).

One such family of PRRs is the Toll-like receptor (TLR) family. Each member of this family of type I transmembrane proteins consists of an extracellular domain with a series of leucine-rich repeats and an intracellular signaling domain that shares homology with the interleukin (IL)-1 receptor, called the Toll/IL-1R (TIR) domain. The first Toll receptor was identified in *Drosophila* where it was implicated in anti-fungal immunity. In particular, Toll was shown to be necessary for activation of Rel/nuclear factor (NF)-κB transcription factors, which in turn led to expression of anti-fungal peptides.

A large family of mammalian TLRs has now been identified. Currently, there are 10 TLRs known in humans. The ligands for these mammalian TLRs are continuing to be defined, but, in general, TLRs recognize PAMPs. TLRs are expressed on DCs and macrophages and control activation of these APCs. Activation of TLRs initiates a signal transduction cascade which eventually leads to activation of NF-κB, up-regulation of costimulatory molecules on APC, and the production of pro-inflammatory cytokines, such as tumor necrosis factor-α, IL-1, and IL-6. Thus, TLRs couple microbial recognition to pro-inflammatory signals and APC activation which leads to T cell priming and initiation of adaptive immune responses.

2 Toll-Like Receptors and Their Ligands

The last several years have seen a steady increase both in the number of ligands identified for TLRs and in the number of TLR family members. There are currently 10 TLRs identified in humans, though ligands have not been identified for all 10. In Table 1 we have listed the known ligands for each of the mammalian TLRs. Collectively, these ligands are extremely diverse, both structurally and in terms of the pathogens from which they are derived. One difficulty in constructing such a table is evaluating the approaches used to demonstrate the recognition of a PAMP by a particular TLR, especially considering the possibility of cross-contamination of microbial TLR ligand preparations leading to false-positive results. This has been a problem that continues to plague the identification of Toll ligands. The use of knockout mice is the most rigorous test for a ligand–receptor pair, although ligands have been identified using other means.

2.1 TLR2 Ligands

TLR2 has the ability to recognize a broad range of microbial products. Ligands for TLR2 include peptidoglycan from Gram-positive bacteria (SCHWANDNER et al. 1999; TAKEUCHI et al. 1999), lipoproteins (diacylated and triacylated) (ALIPRANTIS et al. 1999; BRIGHTBILL et al. 1999; FRENDEUS et al. 2001; TAKEUCHI et al. 2000),

Table 1. Known TLR ligands. The known TLR ligands are listed together with the source of the ligand and the corresponding TLR (see text for discussion and references)

Ligand	Source	TLR
Peptidoglycan	Gram+ bacteria	TLR2
Lipoprotein (triacylated)	Bacteria	TLR2
Lipoprotein (diacylated)	Mycoplasma	TLR2 + TLR6
Zymosan	Yeast	TLR2
Lipoteichoic acid	Gram+ bacteria	TLR2[a]
GPI anchors	*Trypanosoma cruzi*	TLR2
Outer membrane protein A	*Klebsiella pneumoniae*	TLR2
Phosphatidylinositol dimannoside	Mycobacteria	TLR2
Glycolipid lipoarabinomannan	Mycobacteria	TLR2
19kDa lipoprotein	Mycobacteria	TLR2
Phenol soluble modulin	*Staphylococcus epidermidis*	TLR2 + TLR6
Leptospiral LPS	*Leptospira interrogans*	TLR2
Poryphyromonas gingivitis LPS	*Poryphyromonas gingivitis*	TLR2
Poly I:C (dsRNA)	dsRNA viruses	TLR3
Lipopolysaccharide	Gram– bacteria	TLR4
Heat-sensitive, cell-associated factor	Mycobacteria	TLR4
RSV F protein	Virus	TLR4[a]
Hsp60	Human	TLR4[a]
Fibronectin extra domain A	Human	TLR4[a]
Flagellin	Bacteria	TLR5
Unmethylated CpG DNA	Bacteria, viruses?	TLR9

[a] Entries with conflicting data or for which the biological significance is still unclear.

zymosan (a component of yeast cell walls) (UNDERHILL et al. 1999a), glycosylphosphatidylinositol (GPI) anchors from *Trypanosoma cruzi* (CAMPOS et al. 2001), lipoarabinomannan from mycobacteria (MEANS et al. 1999a; UNDERHILL et al. 1999b), a phenol soluble modulin from *Staphylococcus epidermis* (HAJJAR et al. 2001), and lipopolysaccharide (LPS) variants from *Leptospira interrogans* (WERTS et al. 2001) and *Poryphyromonas gingivitis* (HIRSCHFELD et al. 2001). There are conflicting reports regarding the recognition of lipoteichoic acid (LTA). Experiments using knockout mice have implicated both TLR2 and TLR4, individually, as the receptor necessary for LTA responsiveness (MICHELSEN et al. 2001; TAKEUCHI et al. 1999). The reasons for these conflicting results are not clear. In addition, only peptidoglycan from Gram-positive bacteria has been shown to stimulate TLR2. Gram-negative bacteria have a peptidoglycan layer between their inner and outer membranes. Whether this peptidoglycan can stimulate TLR2 has not been formally demonstrated. It is also not clear what structural features make each of these components a viable TLR2 ligand. Apparently, the structural features recognized by TLR2 are quite broad, and this breadth of recognition stems, at least in part, from TLR2's ability to pair with other TLRs (OZINSKY et al. 2000). Both TLR1 and TLR6 have been shown to signal cooperatively with TLR2, and this cooperation can affect the specificity of the receptor pair. For instance, TLR6-deficient mice are not responsive to MALP-2 (mycoplasmal macrophage-activating lipopeptide-2 kDa), a diacylated lipopeptide from mycoplasma, yet respond normally to triacylated bacterial lipopeptides (TAKEUCHI et al. 2001). Interestingly, TLR2-deficient mice are unresponsive to both forms of lipopeptide, indicating that TLR6, when associated with TLR2, discriminates between these closely related PAMPs. In light of these data, it seems likely that TLR1, together with TLR2, recognizes triacylated lipopeptides, although this specificity has not been demonstrated directly.

2.2 TLR4 Ligands

TLR4 is the long sought LPS receptor. It was first identified and characterized independently of this recognition (MEDZHITOV et al. 1997). Subsequently, a mutation in TLR4 was shown to be responsible for the LPS-hyporesponsive phenotype of the C3H/HeJ mouse strain (POLTORAK et al. 1998; QURESHI et al. 1999). Recognition of LPS by TLR4 requires a number of accessory molecules. LPS must be bound by a serum protein called LPS binding protein (19) (WRIGHT et al. 1989). LBP transfers LPS monomers to the GPI-linked protein, CD14. CD14 is essential for TLR4 recognition of LPS as evidenced by the unresponsiveness of CD14-deficient mice (HAZIOT et al. 1996). Finally, a soluble protein called MD-2 is also involved in LPS recognition, although the exact mechanism of MD-2 action is unknown (SHIMAZU et al. 1999). TLR4 has also been implicated in the recognition of LTA (as described above) as well as a heat-sensitive, cell-associated factor from *Mycobacterium tuberculosis* (MEANS et al. 1999b).

In addition to these microbial ligands, two endogenous proteins have been reported to be ligands for TLR4: heat-shock protein 60 (Hsp60) (OHASHI et al.

2000; VABULAS et al. 2001) and the extra domain A of fibronectin (OKAMURA et al. 2001). TLR2 may also be capable of recognizing Hsp60 (VABULAS et al. 2001). Hsp60 is a chaperone that is highly conserved from bacteria to humans. Normally, Hsp60 would not be available for TLR binding, as it is sequestered intracellularly. However, it has been suggested that ligands such as HSPs can be released upon necrotic cell death, a phenomenon which can lead to inflammation. Furthermore, it has recently been shown that necrotic cells can lead to activation of NF-κB in macrophages and fibroblasts via TLRs (LI et al. 2001). Interestingly, this activation led to expression of genes implicated in tissue repair responses. Therefore, it is possible that TLRs are involved in detection of necrotic cell death and that, in this context, TLRs lead to a different gene expression profile than in response to microbial infection. One concern with analysis of endogenous ligands, though, is that they are usually expressed recombinantly. Thus, it remains a possibility that the effects observed are due to contamination with LPS or some other PAMP. While researchers in this area are careful to control for this possibility, it remains a point of concern as long as bacterially expressed proteins are used.

An intriguing ligand identified for TLR4 is the F protein of *Respiratory syncytial virus* (RSV) (KURT-JONES et al. 2000). F protein was shown to induce IL-6 production by macrophages in a TLR4-, CD14-dependent manner. It remains to be seen whether this protein represents a true viral PAMP, especially in view of the fact that it is a protein. Many viruses have evolved to interact with host proteins for their own benefit, either to gain entry into cells or to manipulate the cellular response to a state more conducive for viral replication.

2.3 TLR5 Ligands

The only known ligand for TLR5 is flagellin, a component of bacterial flagella (HAYASHI et al. 2001). Flagellin is a highly conserved protein, presumably due to structural constraints necessary for its function. This conservation makes flagellin a likely target for innate immune recognition, as demonstrated by the recognition of flagellin by PRRs of the innate immune systems of plants (GOMEZ-GOMEZ and BOLLER 2000). Nevertheless, flagellin stands out as the only protein PAMP described to date that does not require post-translational modifications unique to microorganisms.

2.4 TLR9 Ligands

The enigmatic observation that unmethylated CpG motifs present in bacterial DNA (CpG-DNA) are potent stimulators of the immune system has fascinated immunologists for many years (KRIEG 2000; KRIEG et al. 1995). This property of CpG-DNA is particularly amazing in that a single nucleotide substitution can abolish this stimulatory activity. The finding that TLR9-deficient mice were unresponsive to CpG motifs identified TLR9 as the elusive 'CpG DNA' receptor (HEMMI et al. 2000). The mechanism by which TLR9 recognizes CpG-DNA with such specificity remains unclear and is the focus of intense research. It is known that CpG DNA must be internalized to activate TLR9, suggesting that TLR9 may be localized intracellularly (HACKER et al. 1998).

2.5 TLR3 Ligands and Viral Recognition

TLR3 has recently been shown to function as a receptor for double-stranded RNA (dsRNA) (ALEXOPOULOU et al. 2001). dsRNA is the genetic component of an entire class of viruses but can also be produced during replication and expression of genes encoded by other classes of viruses. In general, viruses present a unique challenge to the innate immune system. By replicating within host cells, viruses avoid the generation of any unique metabolic products that could be targeted for innate immune recognition. One exception appears to be dsRNA which can be considered a molecular pattern of viruses. Mechanisms to recognize dsRNA intracellularly are well documented. In particular, the protein kinase PKR is involved in recognition of dsRNA and the subsequent shut-down of translation (WILLIAMS 1999). TLR3 recognizes extracellular dsRNA, which may be important for the detection of viral dsRNA released from infected cells.

An intriguing possibility is that TLR9 has also evolved to recognize viral genomes. The stimulatory CpG-DNA found in bacterial DNA can also be found within viral genomes. This possibility seems more plausible when it is considered that TLR9 is expressed in a subset of DCs called plasmacytoid DCs (KADOWAKI et al. 2001). These cells have been implicated in anti-viral responses, express TLR9, and produce type I interferons (IFNs) when treated with CpG-DNA. Thus, it seems very possible that TLR9 could be involved in responses to viruses. In fact, considering the redundancy in TLR recognition of bacteria, it is more likely that TLR9 evolved to recognize viruses than bacteria.

The issue remains of how TLR3 and TLR9 gain access to viral genomes. As most viruses uncoat their genomes only in the cytosol, it is not immediately obvious why the immune system would target viral genomes using extracellular receptors. It is possible that TLR3 and TLR9 could recognize dsRNA or CpG-DNA released from viruses that are degraded in lysosomes after internalization. Opsonization of viruses may prevent fusion with the membrane and instead lead to eventual degradation. The role of TLRs in viral immunity will not be clear until viral infection experiments are performed with TLR-deficient mice. Until then, we must be satisfied with the observation that dsRNA and CpG-DNA are the most likely molecular signatures of viruses described thus far.

2.6 Additional TLR Ligands

There have not been any ligands found yet for TLR7, TLR8, or TLR10. However, based on the high degree of similarity between the TIR domains of TLRs 1, 6, and 10, it is likely that TLR10 pairs with TLR2. One or more of the known TLR2 ligands may be recognized together with TLR10. That leaves TLR7, TLR8, and any unidentified TLRs as the remaining receptors without ligands. Even without the complete set of TLR ligands, though, several trends can be identified. First, it seems likely that multiple ligands will exist for each TLR. Second, to accommodate the recognition of such structurally diverse ligands, TLRs must have multiple ligand binding sites. Finally, the growing list of ligands seems to be converging on a relatively small set of molecular patterns. This last trend is consistent with the targeting of the few invariant features common to different types of microbes. One prediction from these trends of innate immune recognition is that TLRs do not need to generate a large combinatorial diversity of specificities to achieve broad recognition of microbes. Instead, what is confusing about our current understanding of TLR recognition is the apparent redundancy in recognition of particular microbes.

A major challenge will be to make sense of the recognition of certain classes of microbes by certain TLRs. What is the consequence of recognition by one TLR versus another? There has been much enthusiasm for potential conceptual links between invertebrate and vertebrate innate immunity, especially in light of the apparent discrimination of classes of pathogens by invertebrate immune systems (HOFFMANN et al. 1999). However, it is becoming clear that the mammalian system does not separate microbes into simple categories based on TLR recognition, as was originally suggested based on the *Drosophila* system. The complexity of the system is evident just from the list of known TLR2 ligands, which represent components of Gram-negative bacteria, Gram-positive bacteria, yeast, and protozoa. Discrimination of pathogen class as it occurs in *Drosophila* cannot be achieved when so many different ligands exist for a single receptor. There is also significant redundancy in terms of the ligands present on microbes themselves. Consider that based on our current list of ligands it is highly possible that an individual bacterium could express PAMPs that would stimulate TLRs 1, 2, 4, 5, 6, and 9. It is difficult to envision how specific effector responses to one class of microbe versus another are achieved with such redundancy. Where is the specificity when ligands from such disparate sources are engaged by the same receptor? These observations underscore the fact that we do not yet possess the information to discern the specificity within this system of recognition. There are undoubtedly specific responses associated with each TLR. The challenge will be to link these responses with the pathogens that elicit them and gain some understanding of how the system works to classify different types of infection.

3 Toll-Like Receptor Signaling

Activation of TLRs leads to the expression of a variety of genes involved in the host response to infection, including cytokines, chemokines, and major histocompatibility complex (MHC) and costimulatory molecules. In addition, TLRs can lead to up-regulation of effector mechanisms such as production of reactive oxygen intermediates and antimicrobial peptides. Many of the responses are shared by all TLRs due to the conserved signaling pathways common to TLRs and IL-1R. It is becoming increasingly evident, though, that there are sets of unique genes induced by specific TLR family members.

As suggested by the homology shared between their intracellular TIR domains, the IL-1R and TLR families share a common signaling pathway which leads to the activation of NF-κB as well as mitogen-activated protein kinases (MAPKs). This pathway is initiated through homophilic interactions between the TIR domains of the receptor and MyD88, an essential adaptor in TLR/IL-1R signaling (Burns et al. 1998; Medzhitov et al. 1998; Muzio et al. 1997; Wesche et al. 1997). MyD88 consists of a TIR domain linked to a death domain. The death domain of MyD88 interacts with the death domain of the IL-1R-associated kinase (IRAK), thus recruiting IRAK to the receptor. Recruitment of IRAK is also aided by a second adaptor called TOLLIP (Toll-interacting protein) (Burns et al. 2000). Although TOLLIP has neither a TIR domain nor a death domain, it can associate with IRAK and with the TIR domain of the receptor. The apparent functional redundancy between TOLLIP and MyD88 in terms of IRAK recruitment is not yet understood. Regardless, once recruited to the receptor, IRAK is autophosphorylated and can interact with TRAF6. TRAF6 leads to the activation of TAK1, which, in turn, leads to the activation of NF-κB and the MKK6/p38/JNK pathway (Wang et al. 2001).

The first clue that differences existed in the signaling pathways downstream of individual TLRs came from analysis of MyD88-deficient mice. Macrophages and DCs from these mice did not produce cytokines in response to TLR ligands (Kawai et al. 1999). However, careful analysis of the activation of NF-κB and MAPK pathways in these mice demonstrated that stimulation of MyD88-deficient cells with ligands for TLR4 or TLR3 still resulted in the activation of these signaling pathways, although this activation was delayed compared to wild-type (Alexopoulou et al. 2001; Kaisho et al. 2001; Kawai et al. 1999; Schnare et al. 2000). Even more interesting, though, was the finding that ligands for TLR2 and TLR9 did not lead to MyD88-independent activation of NF-κB or MAPKs. These results demonstrated the existence of at least two different signaling pathways downstream of TLR3 and TLR4, with different dependencies on MyD88, yet the existence of only a MyD88-dependent pathway downstream of TLR2 and TLR9. These pathways were also involved in the maturation of DCs. TLR signaling in DCs leads not only to cytokine production but also to up-regulation of MHC and costimulatory molecules (called DC maturation). DCs derived from MyD88-deficient mice did not produce cytokines when stimulated with LPS or PolyI:C (a synthetic mimic of dsRNA) but did up-regulate MHC and costimulatory molecules (Alexopoulou et al. 2001; Kaisho et al. 2001). This MyD88-independent maturation did not occur in MyD88-deficient DCs treated with ligands for TLR2 or TLR9. Thus, MyD88 is required for cytokine production downstream of all TLRs, but

MyD88-independent signaling events downstream of TLR3 and TLR4 are sufficient for DC maturation.

A clue to understanding these differences in signaling came from the identification of a new adaptor protein called TIRAP (Horng et al. 2001) or Mal (Fitzgerald et al. 2001). TIRAP functions downstream of TLR4 and appears to be responsible for the MyD88-independent signaling observed in response to LPS. TIRAP contains a carboxy terminal TIR domain but lacks a death domain. A dominant-negative form of TIRAP blocks NF-κB activation by TLR4 but not by TLR9 or IL-1R, supporting the role for TIRAP controlling the MyD88-independent pathway (Horng et al. 2001).

Collectively, these results provide evidence that the signaling pathways differ downstream of individual TLRs. In addition to further defining these differences, a next step will be to understand the consequences of differential signaling. There is accumulating evidence that activation of individual TLRs leads to expression of different sets of genes. Recent work demonstrated that the MyD88-independent signaling pathway downstream of TLR4 leads to activation of IFN regulatory factor 3 (IRF-3) and expression of a set of genes, which are not induced by TLR2 (Kawai et al. 2001). Furthermore, studies using gene chip technology have provided data supporting differential gene induction, although these studies tend to use microbes as stimuli instead of individual PAMPs (Huang et al. 2001). A comprehensive analysis comparing gene expression induced by pure TLR ligands has yet to be reported and would be of great interest.

4 Control of Adaptive Immunity by Toll-Like Receptors

More than 10 years ago Janeway proposed a theory of innate immune recognition in which adaptive responses would be linked to recognition of PAMPs by receptors of the innate immune system (Janeway 1989). Expression of TLRs on APCs combined with their specificity for microbial ligands allows them to control adaptive responses. Recent experiments using MyD88-deficient mice have demonstrated the importance of TLRs in generation of adaptive immune responses (Schnare et al. 2001). Mice immunized with protein mixed with Complete Freund's Adjuvant (CFA) generate a strong immune response against the protein. This response is characterized by T-cell proliferation, IFN-γ production by T cells, and the production of antibodies of the IgG2a isotype specific for the protein. In MyD88-deficient mice these responses do not occur. This block is due to the inability of MyD88-deficient mice to respond to the microbial component (*M. tuberculosis*) in CFA. In particular, DCs from MyD88-deficient mice do not mature or produce cytokines in response to *M. tuberculosis*.

Unexpectedly, T helper (Th) type 2 responses in these experiments appeared to be unaffected (Schnare et al. 2001). Antibodies of the IgE and IgG1 isotypes were still raised against the injected protein. This result was surprising in that it suggested that TLR recognition was necessary for priming of Th1 responses but not of Th2 responses. The explanation for this difference may lie in the type of pathogens that elicit each type of response. Generally, Th1 responses are induced by pathogens that would be predicted to have PAMPs, i.e. bacteria, viruses, and protozoa. Th2 responses are raised against worms and other parasites or during allergic responses. Perhaps helminths and allergens lack PAMPs that can be recognized by TLRs and are consequently TLR-independent. The question then becomes how are multicellular parasites recognized by the innate immune system. It is possible that a distinct family of PRRs recognize these types of pathogens and direct the adaptive response accordingly.

5 Conclusions and Perspectives

In the last several years, our understanding of the role that TLRs play in innate immune recognition has continued to increase. It is clear that TLRs play a critical role in immune responses, both innate and adaptive. In the future, research will focus on identification of the complete set of TLR ligands and characterization of the responses associated with activation of a particular TLR. With this information, the challenge will be to develop an understanding of how innate immune recognition of different microbes is linked to specific effector responses and to the direction of adaptive immune responses.

References

Alexopoulou L, Holt AC, Medzhitov R, Flavell RA (2001) Nature 413:732–8
Aliprantis AO, Yang RB, Mark MR, Suggett S, Devaux B, Radolf JD, Klimpel GR, Godowski P, Zychlinsky A (1999) Science 285:736–9
Brightbill HD, Libraty DH, Krutzik SR, Yang RB, Belisle JT, Bleharski JR, Maitland M, Norgard MV, Plevy SE, Smale ST, Brennan PJ, Bloom BR, Godowski PJ, Modlin RL (1999) Science 285:732–6
Burns K, Clatworthy J, Martin L, Martinon F, Plumpton C, Maschera B, Lewis A, Ray K, Tschopp J, Volpe F (2000) Nature Cell Biol 2:346–51
Burns K, Martinon F, Esslinger C, Pahl H, Schneider P, Bodmer JL, Di Marco F, French L, Tschopp J (1998) J Biol Chem 273:12203–9
Campos MA, Almeida IC, Takeuchi O, Akira S, Valente EP, Procopio DO, Travassos LR, Smith JA, Golenbock DT, Gazzinelli RT (2001) J Immunol 167:416–23
Fitzgerald KA, Palsson-McDermott EM, Bowie AG, Jefferies CA, Mansell AS, Brady G, Brint E, Dunne A, Gray P, Harte MT, McMurray D, Smith DE, Sims JE, Bird TA, O'Neill LA (2001) Nature 413:78–83
Frendeus B, Wachtler C, Hedlund M, Fischer H, Samuelsson P, Svensson M, Svanborg C (2001) Mol Microbiol 40:37–51
Gomez-Gomez L, Boller T (2000) Mol Cell 5:1003–11
Hacker H, Mischak H, Miethke T, Liptay S, Schmid R, Sparwasser T, Heeg K, Lipford GB, Wagner H (1998) EMBO J 17:6230–40
Hajjar AM, O'Mahony DS, Ozinsky A, Underhill DM, Aderem A, Klebanoff SJ, Wilson CB (2001) J Immunol 166:15–9

Hayashi F, Smith KD, Ozinsky A, Hawn TR, Yi EC, Goodlett DR, Eng JK, Akira S, Underhill DM, Aderem A (2001) Nature 410:1099–103

Haziot A, Ferrero E, Kontgen F, Hijiya N, Yamamoto S, Silver J, Stewart CL, Goyert SM (1996) Immunity 4:407–14

Hemmi H, Takeuchi O, Kawai T, Kaisho T, Sato S, Sanjo H, Matsumoto M, Hoshino K, Wagner H, Takeda K, Akira S (2000) Nature 408:740–5

Hirschfeld M, Weis JJ, Toshchakov V, Salkowski CA, Cody MJ, Ward DC, Qureshi N, Michalek SM, Vogel SN (2001) Infect Immun 69:1477–82

Hoffmann JA, Kafatos FC, Janeway CA, Ezekowitz RA (1999) Science 284:1313–8

Horng T, Barton GM, Medzhitov R (2001) Nature Immunol 2:835–41

Huang Q, Liu do N, Majewski P, Schulte le AC, Korn JM, Young RA, Lander ES, Hacohen N (2001) Science 294:870–5

Janeway CA Jr (1989) Cold Spring Harb Symp Quant Biol 54 Pt 1:1–13

Kadowaki N, Ho S, Antonenko S, Malefyt RW, Kastelein RA, Bazan F, Liu YJ (2001) J Exp Med 194:863–9

Kaisho T, Takeuchi O, Kawai T, Hoshino K, Akira S (2001) J Immunol 166:5688–94

Kawai T, Adachi O, Ogawa T, Takeda K, Akira S (1999) Immunity 11:115–22

Kawai T, Takeuchi O, Fujita T, Inoue J, Muhlradt PF, Sato S, Hoshino K, Akira S (2001) J Immunol 167:5887–94

Krieg AM (2000) Curr Opin Immunol 12:35–43

Krieg AM, Yi AK, Matson S, Waldschmidt TJ, Bishop GA, Teasdale R, Koretzky GA, Klinman DM (1995) Nature 374:546–9

Kurt-Jones EA, Popova L, Kwinn L, Haynes LM, Jones LP, Tripp RA, Walsh EE, Freeman MW, Golenbock DT, Anderson LJ, Finberg RW (2000) Nature Immunol 1:398–401

Li M, Carpio DF, Zheng Y, Bruzzo P, Singh V, Ouaaz F, Medzhitov RM, Beg AA (2001) J Immunol 166:7128–35

Means TK, Lien E, Yoshimura A, Wang S, Golenbock DT, Fenton MJ (1999a) J Immunol 163:6748–55

Means TK, Wang S, Lien E, Yoshimura A, Golenbock DT, Fenton MJ (1999b) J Immunol 163:3920–7

Medzhitov R, Preston-Hurlburt P, Janeway CA, Jr (1997) Nature 388:394–7

Medzhitov R, Preston-Hurlburt P, Kopp E, Stadlen A, Chen C, Ghosh S, Janeway CA, Jr (1998) Mol Cell 2:253–8

Michelsen KS, Aicher A, Mohaupt M, Hartung T, Dimmeler S, Kirschning CJ, Schumann RR (2001) J Biol Chem 276:25680–6

Muzio M, Ni J, Feng P, Dixit VM (1997) Science 278:1612–5

Ohashi K, Burkart V, Flohe S, Kolb H (2000) J Immunol 164:558–61

Okamura Y, Watari M, Jerud ES, Young DW, Ishizaka ST, Rose J, Chow JC, Strauss JF, 3rd. (2001) J Biol Chem 276:10229–33

Ozinsky A, Underhill DM, Fontenot JD, Hajjar AM, Smith KD, Wilson CB, Schroeder L, Aderem A (2000) Proc Natl Acad Sci USA 97:13766–71

Poltorak A, He X, Smirnova I, Liu MY, Huffel CV, Du X, Birdwell D, Alejos E, Silva M, Galanos C, Freudenberg M, Ricciardi-Castagnoli P, Layton B, Beutler B (1998) Science 282:2085–8

Qureshi ST, Lariviere L, Leveque G, Clermont S, Moore KJ, Gros P, Malo D (1999) J Exp Med 189: 615–25

Schnare M, Barton GM, Holt AC, Takeda K, Akira S, Medzhitov R, (2001) Nature Immunol 2:947–50

Schnare M, Holt AC, Takeda K, Akira S, Medzhitov R (2000) Curr Biol 10:1139–42

Schwandner R, Dziarski R, Wesche H, Rothe M, Kirschning CJ (1999) J Biol Chem 274:17406–9

Shimazu R, Akashi S, Ogata H, Nagai Y, Fukudome K, Miyake K, Kimoto M (1999) J Exp Med 189:1777–82

Takeuchi O, Hoshino K, Kawai T, Sanjo H, Takada H, Ogawa T, Takeda K, Akira S (1999) Immunity 11:443–51

Takeuchi O, Kaufmann A, Grote K, Kawai T, Hoshino K, Morr M, Muhlradt PF, Akira S (2000) J Immunol 164:554–7

Takeuchi O, Kawai T, Muhlradt P, Morr M, Radolf JD, Zychlinsky A, Takeda K, Akira S (2001) Int Immunol 13:933–940

Underhill DM, Ozinsky A, Hajjar AM, Stevens A, Wilson CB, Bassetti M, Aderem A (1999a) Nature 401:811–5

Underhill DM, Ozinsky A, Smith KD, Aderem A (1999b) Proc Natl Acad Sci USA 96:14459–63

Vabulas RM, Ahmad-Nejad P, da Costa C, Miethke T, Kirschning CJ, Hacker H, Wagner H (2001) J Biol Chem 276:31332–9

Wang C, Deng L, Hong M, Akkaraju GR, Inoue J, Chen ZJ (2001) Nature 412:346–51
Werts C, Tapping RI, Mathison JC, Chuang TH, Kravchenko V, Saint Girons I, Haake DA, Godowski PJ, Hayashi F, Ozinsky A, Underhill DM, Kirschning CJ, Wagner H, Aderem A, Tobias PS, Ulevitch RJ (2001) Nature Immunol 2:346–52
Wesche H, Henzel WJ, Shillinglaw W, Li S, Cao Z (1997) Immunity 7:837–47
Williams BR (1999) Oncogene 18:6112–20
Wright SD, Tobias PS, Ulevitch RJ, Ramos RA (1989) J Exp Med 170:1231–41

Toll-Like Receptor-5 and the Innate Immune Response to Bacterial Flagellin

K.D. SMITH* and A. OZINSKY

The innate immune system identifies the presence of infection by detecting structures that are unique to microbes and that are not expressed in the host. The bacterial flagellum (Latin, a whip) confers motility on a wide range of bacterial species. Vertebrates, plants, and invertebrates all have evolved flagellar recognition systems that are activated by flagellin, the major component of the bacterial flagellar filament. In mammals, flagellin is recognized by Toll-like receptor-5 and activates defense responses both systemically and at epithelial surfaces. Here, we review the role for Toll-like receptor-5 in mediating the mammalian innate immune response to flagellin, and how this provides for defense against infections caused by many different species of flagellated bacteria.

1 Introduction

The innate immune system rapidly mobilizes defenses to respond effectively to threats posed by a multitude of different microbes. The discovery of the Toll-like receptor (TLR) family has provided great insight into how the innate immune

Institute for Systems Biology, 1441 North 34th Street, Seattle, WA 98103, USA
*Department of Pathology, University of Washington, Box 35610 1959 NE Pacific Street, Seattle, WA 98195, USA

system recognizes microbes and initiates inflammatory signaling. Receptors of the innate immune system have been selected to identify microbial components that are distinctly non-self (MEDZHITOV and JANEWAY 1997). The TLR family of innate immune receptors is utilized by a vast array of multicellular eukaryotic organisms, ranging from insects to plants to humans (ADEREM and ULEVITCH 2000; MEDZHITOV 2001). Toll-like receptors detect a wide range of microbial products, including lipopolysaccharide (LPS), peptidoglycan, yeast cell wall, lipoteichoic acid, bacterial lipoproteins, bacterial heat shock proteins, mycobacterial phosphoinositol mannosides, bacterial CpG-DNA, double-stranded RNA, and bacterial flagellin (UNDERHILL and OZINSKY 2002). Once engaged, these receptors activate host cells, initiate local and systemic inflammatory responses, and drive the differentiation of antigen presenting cells to promote an adaptive immune response (AKIRA et al. 2001; MEDZHITOV 2001).

Over the past few years, it has been determined that bacterial flagella activate host innate immune defenses, both in white blood cells (monocytes and macrophages), and in epithelial cells. Recently, we have determined that TLR5 mediates the innate immune response to flagella through recognition of flagellin, the major constituent of the bacterial flagellar filament (HAYASHI et al. 2001). In this review, we will discuss how TLR5 activation by flagellin may mediate innate immune defense against many different species of flagellated bacteria.

2 Bacterial Flagellin

Flagella are responsible for bacterial motility and are regulated to produce directed movement under the control of sensors that detect environmental cues. It is presumed that the ability to approach nutrients and to avoid noxious surroundings confers a strong survival advantage that has led to the evolutionary conservation of structures required for flagellar assembly in many different bacterial species. The concerted action of more than 50 genes is required for eubacterial flagellar regulation and assembly (CHILCOTT and HUGHES 2000). Flagella are composed of three major components: a rotary motor (the basal body complex) that is inserted in a ring structure in the peptidoglycan and LPS cell wall layers, a flexible universal joint (the hook) and a whip-like filament (Fig. 1). The hook and filament are extracellular structures. The motor converts chemical energy into the mechanical force that rotates the hook. The hook is an approximately 55nm long flexible structure that transmits torque to the filament (HIRANO et al. 1994). The filament, which can be up to 15μm long, is made almost entirely from polymerized flagellin subunits. The filament is assembled with 11 flagellin monomers being stacked per two turns like bricks forming a spiral staircase-like cylinder. The filament also has a cap complex that functions as a chaperone to facilitate filament assembly (Fig. 1). The ability of the flagellum to generate motility depends on the direction of rotation of the flagellar motor, which, by switching its rotation in

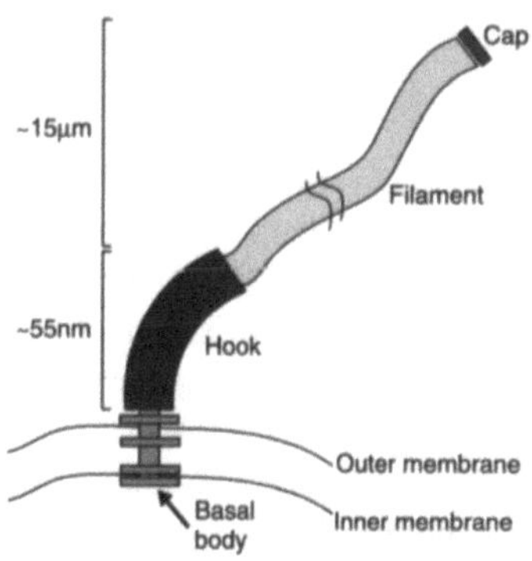

Fig. 1. A model of the bacterial flagellum. The basal body is embedded in the bacterial membranes and is attached to the hook, to form the basal body–hook complex. Flagellin is secreted through this complex to form the filament with the assistance of the cap

response to environmental changes, causes beneficial stimuli to evoke a directed, coordinated movement (running), and noxious stimuli to evoke a non-directed, tumbling state. The structural basis for effects result from two different packing interactions between the flagellin monomers in the filament that cause the filament to switch between left- and right-handed supercoiled forms (Samatey et al. 2001). The flagellin protein has three globular domains (Fig. 2) (Samatey et al. 2001). The D1 domain forms the inner core of the flagellar filament. This domain consists of evolutionarily conserved α-helical regions of the amino and carboxyl termini, and is required for assembly of the filament. The D2 and D3 domains are encoded by the central region of flagellin gene, and are less well conserved in different bacteria. They make up most of the external surface of the flagellar filament, which is a common target of host antibody responses, and form the molecular determinants of the so-called H-antigen, that is widely used for subtyping bacterial strains. Evasion of antibody-mediated immune responses is believed to drive antigenic variation in the D2 and D3 domains. In fact the D3 domain is not required for motility, and is not present in the flagellin molecules of some bacteria.

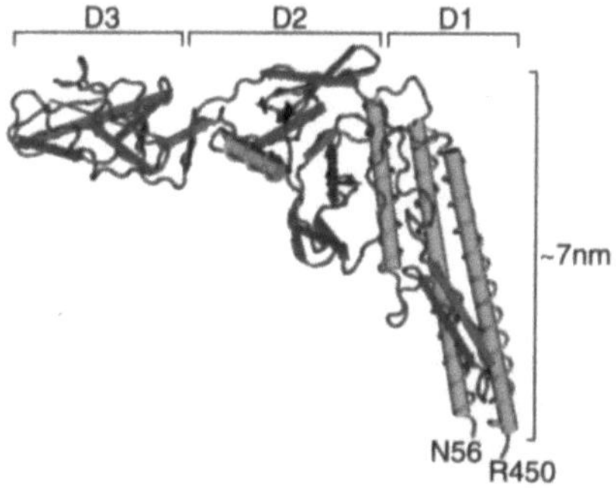

Fig. 2. Crystal structure of flagellin from *Salmonella typhimurium*. Flagellin has three domains, and is oriented with the D1 domain lying innermost in the filament, and the D2 and D3 domains forming the exposed surface of the filament. The amino and carboxy termini of flagellin are not shown. Figure derived from the Genbank Protein Structure Data Bank (accession code 1I01) (Samatey et al. 2001)

3 Toll-Like Receptor-5

The gene for TLR5 is composed of a single exon on human chromosome 1q41-1q42 (CHAUDHARY et al. 1998; ROCK et al. 1998) and encodes a protein of 858 amino acids (Fig. 1). TLR5 has an extracellular domain that is composed of a membrane-distal leucine-rich repeat (LRR) domain and a membrane-proximal LRR C-terminal domain (a cysteine-rich domain often found at the carboxyl terminal end of LRRs). TLR5 has a series of 15 imperfect LRRs in the extracellular domain (Fig. 3). There are six predicted N-glycosylation sites in the external domain of human TLR5, and five of these sites are conserved in the murine molecule. TLR5 has a single transmembrane domain, and a cytoplasmic tail that contains a Toll/interleukin (IL)-1 receptor (TIR) homology domain. The TIR domain is present in all Toll family members, the IL-1 receptor family, the IL-18 receptor, and a subset of plant resistance gene family members (DANGL and JONES 2001; SIMS 2002). Thus, the TIR domain controls an evolutionarily conserved host defense function in plants, insects and animals (AKIRA et al. 2001; DANGL and JONES 2001; IMLER and HOFFMANN 2001). Within the cytoplasmic domain there are several potential sites for phosphorylation (Fig. 3). A potential tyrosine phosphorylation site (residues 713–719) is unique to TLR5, as is a potential casein kinase II phosphorylation site (residues 686–689). There are also three potential protein kinase C phosphorylation sites (residues 700–702, 751–753 and 805–807).

4 TLR5 Recognition of Flagellin

Recently, we have determined that mammalian cells utilize TLR5 to mediate inflammatory responses to flagellin (HAYASHI et al. 2001). In these studies we used a TLR5-dependent reporter system to detect stimulatory activity present in a variety of bacterial cultures. This activity was purified and identified as flagellin by tandem

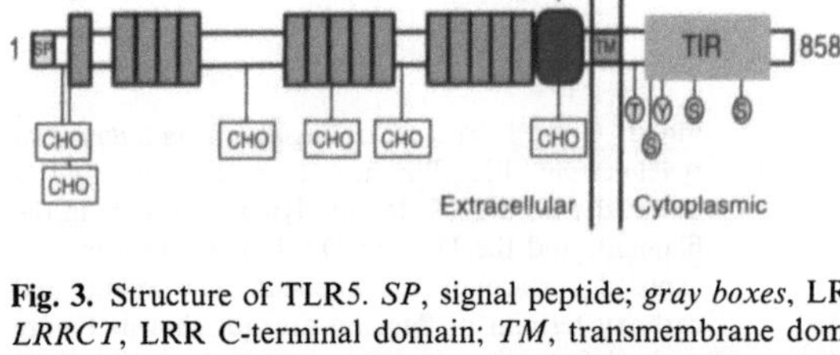

Fig. 3. Structure of TLR5. *SP*, signal peptide; *gray boxes*, LRRs; *CHO*, potential N-glycosylation site; *LRRCT*, LRR C-terminal domain; *TM*, transmembrane domain; *TIR*, Toll/IL-1R homology domain; *T*, potential casein kinase II phosphorylation site; *S*, potential protein kinase C phosphorylation site; *Y*, potential tyrosine kinase phosphorylation site

mass spectrometry. Human and murine TLR5 recognizes bacterial flagellin from both Gram-positive and Gram-negative bacteria, and activates the transcription factor, nuclear factor (NF)-κB, leading to tumor necrosis factor (TNF)-α production. Expression of *Listeria monocytogenes* flagellin in non-flagellated *Eschericia coli* conferred the ability to activate TLR5, whereas deletion of the flagellin genes from *Salmonella typhimurium* abrogated TLR5-stimulating activity. The data indicate that TLR5 permits mammals to specifically detect a diverse set of flagellated bacterial pathogens.

The role for TLR5 in inflammatory responses to flagellin is supported by the findings that MyD88 and IL-1R associated kinase (IRAK) (Fig. 4), components required for signaling by other TLRs, also are required for the flagellin response (Hayashi et al. 2001; Moors et al. 2001). We injected flagellin into the peritoneum of mice, and demonstrated that the systemic IL-6 response was absent in animals lacking the signal adapter, MyD88 (Hayashi et al. 2001). In the human THP-1 promonocytic cell line, flagellin activates the serine/threonine kinase, IRAK, and induces TNF-α (Moors et al. 2001). Recently, Racl and phosphatidylinositol-3 kinase were implicated in TLR2 signaling (Ifn et al. 2000) and their role in TLR5 signaling remains to be examined. In the mouse, the newly discovered adapter, TIRAP/MAL (Fitzgerald et al. 2001; Horng et al. 2001), mediates a MyD88-independent pathway activated by TLR4 (but not by TLR9 or TLR2) that leads to dendritic cell maturation (Kaisho et al. 2001). Future studies will determine whether this molecule participates in TLR5 signaling.

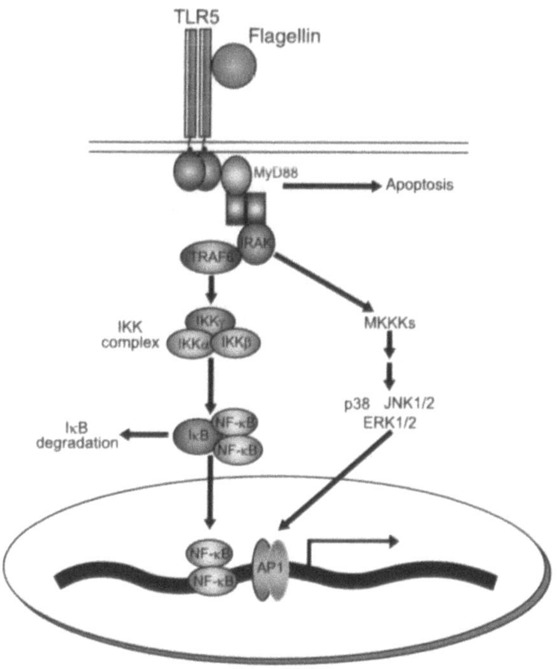

Fig. 4. A postulated minimal model of signal transduction for TLR5. Briefly, TLR5 recruits the adapter protein, MyD88, and the serine/threonine kinase, IRAK, which activates the ring-domain protein, TRAF6. TRAF6 stimulates the IκB kinase complex (*IKK*) that mobilizes the transcription factor, NF-κB, leading to the transcription of inflammatory genes. TRAF6 also activates the mitogen-activated protein kinase pathway (*MKKK*), ultimately leading to the activation of inflammatory genes via the transcription factors that bind to an AP1 site. MyD88 is also known to activate apoptosis

Evidence from a variety of sources indicates that flagellin monomers are able to stimulate TLR5, and that the activity does not require flagellin to be in a polymerized state. Our purification strategy for flagellin involves denaturation by both heat and acid, followed by ultrafiltration through a 100-kDa cut-off filter, which would retain polymerized flagellin (HAYASHI et al. 2001). In epithelial cells, flagellin monomers (as assessed by gel filtration chromatography) were active in inducing IL-8 secretion (GEWIRTZ et al. 2001b). Furthermore, flagellin monomers, which a *Salmonella* strain lacking the flagellar cap protein ($FliD^-$ mutant) secretes, are able to stimulate cytokine production in human monocytes (CIACCI-WOOLWINE et al. 1998).

5 Tissue and Cellular Distribution of TLR5

Several studies have addressed the set of cell and tissue types that express TLR5. In an extensive analysis of human TLR expression by real-time PCR, mRNA for TLR5 was expressed in all of the tissues examined, with the highest level of expression in the ovary, lung, prostate, pancreas, placenta and testis (ZAREMBER and GODOWSKI 2002). TLR5 mRNA also was detected in the kidney, small intestine, colon, spleen, liver, brain, heart, and thymus. The expression of TLR5 in the liver, lung and ovary has been observed by Northern analysis (ROCK et al. 1998; SEBASTIANI et al. 2000). In these tissues, the specific cells that express TLR5 have not been identified. Epithelial cell lines derived from a variety of tissues (lung, intestine, and kidney) have been demonstrated to respond to flagellin, suggesting that epithelia are a source of TLR5 mRNA. In human peripheral blood lymphocytes, TLR5 is predominantly expressed by $CD14^+$ cells, with 10–100-fold lower mRNA levels detected in B and T cells (ZAREMBER and GODOWSKI 2002). Similar findings were obtained by Northern analysis, where TLR5 mRNA is detected in myelomonocytic cells, but is absent (or barely detectable) in lymphoid cells and natural killer cells (MUZIO et al. 2000). In dendritic cells, TLR5 is expressed only in the myeloid subset (that also express TLR2, TLR3, TLR4, TLR6, TLR8) and not in the plasmacytoid subset (that express only TLR7 and TLR9) (JARROSSAY et al. 2001). TLR5 also is expressed in freshly isolated human monocytes and in immature myeloid dendritic cells, but not in plasmacytoid pre-dendritic cells (KADOWAKI et al. 2001). The elements that regulate cell type expression of TLR5 have not been determined. Likewise, the responses of most of these cell and tissue types to flagellin have not been examined in vivo or in isolated primary cells.

6 Innate Immune Response to Flagellin

Bacterial flagellins are potent immunogens that evoke humoral and cellular immune responses. The capacity of flagellin to induce an adaptive immune response has been known for a long time, and it is only recently that the innate response to flagellin has been appreciated in its molecular detail. The potent activity of flagellin to provoke an antibody response played a key role in many of the seminal discoveries, made in the 1960s, by Ada, Nossal and others on the mechanisms of antibody production and, indeed, in proving correct key features of the clonal selection theory of acquired immunity (ADA and BYRT 1969).

Bacterial flagellin fulfills many of the features of a pathogen-associated molecular pattern (PAMP) (for discussion, see MEDZHITOV 2001). Flagellin is expressed as a highly repetitive pattern on the surface of flagellated bacteria. Bacterial flagella, as mentioned above, provide a vital function for bacterial survival. Because flagellin's TLR5-stimulatory activity is derived entirely from its polypeptide structure, and does not require additional microbial-specific chemical modifications (K.D. Smith et al., unpublished results), flagellin also is unique amongst other PAMPs derived from the microbial cell wall. Higher eukaryotes that strive to protect themselves from bacterial infections, such as plants, insects or vertebrates, do not posses proteins with significant homology to flagellin. The strong survival advantage imparted by motility has presumably imposed structural constraints on flagellin. These constraints, in turn, have been targeted by the innate immune system.

6.1 Monocytes/Macrophages

The innate immune response to flagellin is elicited systemically, and at epithelial surfaces in a variety of organs. In mice, intravenous or intraperitoneal administration of flagellin causes a systemic response that has many features in common with the response to other microbial products (such as LPS), and may lead to hypotension, shock and even death (EAVES-PYLES et al. 2001). During Gram-negative sepsis, flagellin is present systemically in animals and even was measured at concentrations of 0.4–2μg/ml in the serum of rats during sepsis (EAVES-PYLES et al. 2001). In mice, intravenous bacterial flagellin is a potent activator of inflammatory responses, and the secretion of cytokines such as TNF-α, IL-6, IL-12p40, IL-10, macrophage inflammatory protein (MIP)-1α and interferon (IFN)-γ have been reported, as well as the production of reactive nitrogen intermediates (EAVES-PYLES et al. 2001; LIAUDET et al. 2002). We have observed the induction of IL-6 in C57BL/6 mice when flagellin is injected into the peritoneum, in the absence of prominent TNF-α, IL-1 or IL-12 production (HAYASHI et al. 2001; K.D. Smith et al., unpublished results). While the magnitude of cytokine induction by flagellin is significantly lower than for LPS, flagellin may be an important modulator of Gram-negative shock, and amplification of cytokine responses occurs when LPS and flagellin are co-administered (LIAUDET et al. 2002). In mice, the cells that generate the systemic flagellin response have not been specifically identified. In humans, primary monocytes, monocyte-derived macrophages, and a human promonocytic cell line (THP-1) produce TNF-α, IL-1β, IL-6, IL-10 and IFN-γ in response to flagellin purified from several Gram-negative species (CIACCI-WOOLWINE et al.

1998, 1999; McDermott et al. 2000; Wyant et al. 1999). Although formally not yet demonstrated, it is presumed that human dendritic cells also respond to flagellin.

6.2 Epithelial Cells

The mechanism by which epithelial cells detect flagella has been intensively studied over the past few years due to the suspicion that these pathways might be activated by infections initiated at epithelial surfaces, and might participate in inflammatory diseases such as ulcerative colitis and Crohn's disease. In addition to participating in innate defense by forming a barrier to infection, epithelial cells respond to bacteria by secreting the chemokines, IL-8, (important for recruiting neutrophils) (Steiner et al. 2000) and CCL20 (important for recruiting dendritic cells) (Sierro et al. 2001) in response to flagellin. Other epithelial cell responses to flagellin include synthesis of inducible nitric oxide synthase and β-defensin-2 (Ogushi et al. 2001). As in macrophages, activation of epithelial cells by flagellin involves degradation of IκB and mobilization of the transcription factor, NF-κB (Ogushi et al. 2001).

Some differences exist between how epithelial cells and monocyte/macrophages recognize flagellin. Intestinal epithelial cells are polarized. Their apical membrane is exposed to the intestinal lumen, and their basolateral membrane contacts adjacent cells and the underlying basement membrane. Interestingly, flagellin does not activate a cytokine response when applied to the apical (lumenal) surface of epithelial cells (Gewirtz et al. 2001b). Activation occurs by a two-step process. First, bacterial–epithelial cell interaction induces flagellin translocation to the basolateral surface, and, second, basolaterally expressed TLR5 recognizes flagellin and elicits IL-8 production (Gewirtz et al. 2001a,b). The mechanism by which flagellin is translocated across epithelial cells is not known. Although flagellated commensal bacteria do not mediate flagellin translocation, and soluble flagellin is not translocated, certain flagellated bacterial pathogens (such as *S. typhimurium*) translocate flagellin efficiently (Gewirtz et al. 2001b). Translocation (and IL-8 secretion) does not require bacterial invasion of the epithelial cells (Eaves-Pyles et al. 1999; Gewirtz et al. 1999; Gewirtz et al. 2001b). The selective recognition of flagellin at the basolateral surface is accounted for by the expression of TLR5 on the basolateral surface of polarized epithelial cells and its absence from the apical surface (Gewirtz et al. 2001a). The cytoplasmic tail of TLR5 contains potential tyrosine-based and dileucine-based sorting signals, which have been implicated in the basolateral sorting of other proteins (Mostov et al. 2000). However, the mechanism that targets TLR5 to the basolateral surface of polarized epithelia has not been determined.

The polarized expression of TLR5 on epithelial cells enables epithelia to tolerate the heavy commensal bacterial load in the intestinal lumen while preserving the capacity to respond to pathogens that penetrate through this layer. Excluding TLR5 from the lumenal face of the mucosa permits the innate immune system to target a structure that is widely expressed on commensal organisms as well as pathogens. TLR5 is well poised to handle these invading pathogens, as it recognizes the flagellar structures that are required for tissue invasiveness by many bacteria (DIBB-FULLER et al. 1999; FLEISZIG et al. 2001; GRANT et al. 1993). In this scheme, the innate immune system does not distinguish between commensal and pathogenic organisms. Any flagellated organism that penetrates the epithelial layer will be considered a threat and subjected to TLR5-mediated host defense.

This hypothesis has features in common with the mechanism of detection of intracellular LPS by epithelial cells. Epithelial cells respond to LPS when it is delivered to intracellular sites by invasive organisms (such as *Shigella flexneri*) or experimentally (by microinjection of LPS). Under these circumstances, LPS activates the intracellular CARD4/NOD1 detection system (GIRARDIN et al. 2001). Thus, although lumenal organisms are not stimulatory, invasive organisms can be detected by the intracellular LPS recognition system, or by the basolateral flagellin detection system.

7 Infections Caused by Flagellated Bacteria

Many pathogenic bacteria are flagellated, and the expression of flagella has been shown to be required for the virulence of several different bacterial species. These include mucosal pathogens that infect the gastrointestinal tract, the urinary tract, and the respiratory tract (Table 1).

Flagellated bacteria are the most common cause of acute food borne intestinal illnesses, which affects 5–6 million people annually in the USA. Multidrug-resistant *S. typhimurium* is an emerging food borne pathogen (HUMPHREY 2001). In a calf model of enterocolitis, the production of functional flagella by *S. typhimurium* is required for tissue invasiveness and the recruitment of protective neutrophils

Table 1. Diseases caused by the flagellated bacteria discussed in this review

Bacterium	Disease
Salmonella typhimurium	Enteritis
Campylobacter jejuni	,,
Enteroaggregative *Escherichia coli*	,,
Vibrio cholerae	,,
Helicobacter pylori	Gastric ulcers
Escherichia coli	Urinary tract infections
Proteus mirabilis	,,
Serratia marcescens	,,
Legionella pneumophila	Pneumonia
Pseudomonas aeruginosa	,,
Treponema pallidum	Syphilis
Borrelia burgdorferi	Lyme disease
Listeria monocytogenes	Listeriosis

(Schmitt et al. 2001). Murine TLR5 lies within the chromosomal site harboring a quantitative trait locus for susceptibility to *Salmonella* infections in the MOLF/Ei mouse strain (Sebastiani et al. 2000). In addition, TLR5 mRNA expression in the liver of susceptible MOLF/Ei mice is fourfold lower than in the resistant C57BL/6J or the 129/Sv murine strains. These data suggest that TLR5 plays a role in murine susceptibility to *Salmonella* infections. *Campylobacter* is a major cause of acute colitis, and is associated with chronic sequelae such as Guillain-Barré syndrome and reactive arthritis. At least one study of Guillain-Barré syndrome suggests that a polypeptide sequence within the *Campylobacter* flagellin protein is a marker for disease (Tsang et al. 2001). In addition, flagellated organisms cause a broad range of serious gastrointestinal infections, including cholera, typhoid, dysentery, and a hemolytic uremic syndrome associated with enterohemorrhagic *E. coli*. *Helicobacter pylori*, the major cause of gastric ulcer disease and gastric cancer worldwide, also is a flagellated organism, and the expression of flagella is required for its ability to penetrate and persist in the gastric mucous layer (Andrutis et al. 1997; Eaton et al. 1996).

Flagellated bacteria (*E. coli*, *Proteus mirabilis* and *Serratia marcescens*) cause the majority of urinary tract infections. For *Proteus mirabilis*, the expression of flagella contributes significantly to ascending urinary tract infections (Mobley et al. 1996). A virulence factor (MrpJ) that co-ordinates the expression of both fimbriae (required for adherence) and flagella (required for motility) has been identified (Li et al. 2001), as well as a homologue that is expressed in *E. coli*.

A broad range of pathogens causes respiratory tract infections, and some flagellated bacteria affect specific patient populations. *Legionella pneumophila* (causative agent of Legionnaires' disease) infections can be severe in elderly and immuno-compromised patients, and *Pseudomonas* spp. commonly infect cystic fibrosis patients. The ability of *L. pneumophila* to invade and infect human promyelocytic HL-60 cells is facilitated by flagella (Dietrich et al. 2001). Similarly, aflagellar *P. aeruginosa* mutants have reduced pathogenicity in a murine pneumonia model (Feldman et al. 1998). Flagella from *P. aeruginosa* induce pulmonary epithelial cells to produce IL-8 and matrilysin, a matrix metalloproteinase that activates host α-defensins (DiMango et al. 1995; Lopez-Boado et al. 2001). Matrilysin over-expression has been implicated in the lung pathology of cystic fibrosis patients (Lopez-Boado et al. 2001).

Other flagellated pathogens include the spirochetes *Treponema pallidum* (causative agent of syphilis) and *Borrelia burgdorferi* (causative agent of Lyme disease). The flagella of these spirochetes lie within the periplasmic space beneath the bacterial outer membrane and therefore are not exposed to the environment. This positioning is thought to be an essential component of the corkscrew form of motility that enables spirochetes to move through viscid medium. Recognition of these types of flagellin by TLR5 has not been demonstrated to date, and if this did occur under physiologic conditions, presumably phagocytosis and unmasking of the outer membrane sheath would precede TLR5 recognition.

8 Bacterial Evasion of TLR5 Recognition

Bacteria have developed several mechanisms to regulate the expression, post-translational modification, and compartmentalization of flagella. Many of these strategies probably serve important physiologic functions, but they also have been implicated in the subversion of antibody responses. Several bacterial species modify flagellin by glycosylation (Brimer and Montie 1998) or methylation of lysine residues (Ambler and Rees 1959). We have expressed flagellin in eukaryotic cells, and determined that bacteria-specific post-translational modifications are not required for flagellin to stimulate TLR5 (K.D. Smith et al., unpublished results). Some bacteria (including *Vibrio*, *Helicobacter* and spirochetes) produce periplasmic or ensheathed flagella that are hidden from direct exposure to the environment and presumably also are hidden from detection by TLR5. The human pathogen, *L. monocytogenes*, regulates flagellar expression in response to temperature and expresses flagellin at ambient temperatures while suppressing flagellar expression at 37°C in the host (Peel et al. 1988). Interestingly, the related, but non-pathogenic species, *L. ivanovii*, does not suppress flagellar expression at 37°C (Katharíou et al. 1995). Some bacteria have multiple homologous flagellin genes and vary the isoforms used to produce flagella in a process termed 'phase variation'. For example, *Vibrio cholera* has five flagellin genes that are approximately 60% homologous to each other (Klose and Mekalanos 1998). The multiple flagellin genes may be a mechanism of antigenic switching to evade antibody responses to flagellin. Flagellar phase variation contributes to *S. typhimurium* virulence in a murine systemic infection model but does not influence invasion of epithelial cells or *Salmonella*-induced enteritis (Ikeda et al. 2001). We have not found a difference in the TLR5-stimulatory activity of FliC and FljB (K.D. Smith et al., unpublished results), the two forms of flagellin in *Salmonella typhimurium*. The role for phase variation in evading TLR5-mediated recognition of other bacteria has not been examined.

9 Recognition of Flagellin by Insects and Plants

Vertebrates, plants, and invertebrates all have independently evolved flagellin recognition systems. In *Drosophila*, flagellar expression by *Bacillus thuringiensis* facilitates bacterial binding to insect cells (Zhang et al. 1995). Flagellin activates defense pathways, and induces the production of the antimicrobial peptide, cecropin (Samakovlis et al. 1992). The use of chemical inhibitors suggests that the pathway involves a G protein (and not protein kinase C) (Samakovlis et al. 1992). The mechanism for recognition of flagellin in *Drosophila* has not been assigned to either the Toll or the Imd pathways that mediate, respectively, the responses to Gram-positive and Gram-negative bacteria (Imler and Hoffmann 2001; Michel et al. 2001). It is important to note that detection pathways that lie upstream of Toll and Imd recognize and distinguish between microbial components (MICHEL et al. 2001). The receptor that confers recognition of flagellin in *Drosophila* has not been identified, and none of the nine TLRs in *Drosophila* (TAUSZIG et al. 2000) is an obvious homologue of mammalian TLR5.

Bacterial flagellin acts as potent elicitor of systemic defense responses in different plant species, including *Arabidopsis*, tomato and rice (FELIX et al. 1999). Genetic mapping in *Arabidopsis* identified two distinct loci, termed *FLS1* and *FLS2*, which are essential for the perception of flagellin (GOMEZ-GOMEZ and BOLLER 2000). The protein encoded by FLS1 has not been identified. FLS2 encodes a transmembrane receptor with an extracellular domain composed of 28 LRRs, and an intracellular serine/threonine protein kinase domain (GOMEZ-GOMEZ and BOLLER 2000). Both the FLS2 extracellular LRR domain and the cytoplasmic kinase domain are required for flagellin recognition and signal transduction (GOMEZ-GOMEZ et al. 2001). Furthermore, FLS1 also participates in flagellin binding to plant cells (BAUER et al. 2001; GOMEZ-GOMEZ et al. 2001). FLS1/FLS2 recognizes a domain that is highly conserved in the amino termini of the flagellins of many different bacterial species. Flg22, a peptide corresponding to this region of *Pseudomonas aeruginosa* flagellin, acts as a potent elicitor of defense responses in plants (FELIX et al. 1999). In tomato cells and membrane preparations, Flg22 interacts with a single high-affinity receptor site formed by FLS1/FLS2 (BAUER et al. 2001; FELIX et al. 1999; MEINDL et al. 2000). Apart from sharing LRRS, FLS2 and TLR5 do not show significant homology, and indeed, the Flg22 peptide does not activate TLR5 (K.D. Smith et al., unpublished results). Interestingly, the plant flagellin detection system fails to recognize the flagellins of the symbionts, *Rhizobium* and *Agrobacterium* (FELIX et al. 1999). In *Agrobacterium*, flagella are required for infectivity (CHESNOKOVA et al. 1997), and recognition of its flagellin would presumably be incompatible with symbiosis. The amino acid sequences of *Rhizobium* and *Agrobacterium* flagellins diverge from non-symbiotic bacteria in the highly-conserved region recognized by FLS1/FLS2 (FELIX et al. 1999), and peptides corresponding to this region of *Rhizobium* and *Agrobacterium* flagellin are not stimulatory to plant cells (FELIX et al. 1999). These data suggest that pressures for symbiosis selected for non-stimulatory forms of flagellin in *Rhizobium* and *Agrobacterium*. Thus, the perception of symbionts in plants differs from the perception of flagellin by TLR5, which has not been shown to distinguish between the flagellins of pathogens and commensal organisms.

10 Conclusion

The definition of flagellin as an activator of the innate immune system, and TLR5 as its receptor, provides a molecular framework for understanding an important part of host–microbe interactions. Currently, these findings are of obvious relevance to immune responses against flagellated bacteria and particularly those generated at epithelial surfaces. The initial in vitro studies have provided a simple model for flagellin detection by TLR5. The precise role for TLR5 in host defense in vivo against specific flagellated bacteria remains to be defined. The contribution of TLR5 to immune defenses will be better understood by a fuller examination of which cells respond to flagellin, of which defenses are dependent on TLR5 activation, and of how these cellular responses differ from those induced by other microbial stimuli. Future studies should provide a clearer understanding of the role for TLR5-mediated responses to flagellin in both acute infections and chronic inflammatory disorders.

Acknowledgements. We acknowledge support from the NIH. We thank D. Underhill, F. Hayashi, P.-Y. Bochud, T. Flo and B. Gantner for helpful discussions and comments on the manuscript, and A. Aderem for his guidance and generous support.

References

Ada GL, Byrt P (1969) Specific inactivation of antigen-reactive cells with ^{125}I-labelled antigen. Nature 222:1291–1292

Aderem A, Ulevitch RJ (2000) Toll-like receptors in the induction of the innate immune response. Nature 406:782–787

Akira S, Takeda K, Kaisho T (2001) Toll-like receptors: critical proteins linking innate and acquired immunity. Nature Immunol 2:675–680

Ambler RP, Rees MW (1959) N-methyl-lysine in bacterial flagellar protein. Nature 184:56–57

Andrutis KA, Fox JG, Schauer DB, Marini RP, Li X, Yan L, Josenhans C, Suerbaum S (1997) Infection of the ferret stomach by isogenic flagellar mutant strains of *Helicobacter mustelae*. Infect Immun 65:1962–1966

IFN L, Mira JP, Teusch N, Kline L, Guha M, Mackman N, Godowski PJ, Ulevitch RJ, Knaus UG (2000) Toll-like receptor 2-mediated NF-κB activation requires a Rac1- dependent pathway. Nature Immunol 1:533–540

Bauer Z, Gomez-Gomez L, Boller T, Felix G (2001) Sensitivity of different ecotypes and mutants of *Arabidopsis thaliana* towards the bacterial elicitor flagellin correlates with the presence of receptor binding sites. J Biol Chem 19:19

Brimer CD, Montie TC (1998) Cloning and comparison of fliC genes and identification of glycosylation in the flagellin of *Pseudomonas aeruginosa* a-type strains. J Bacteriol 180:3209–3217

Chaudhary PM, Ferguson C, Nguyen V, Nguyen O, Massa HF, Eby M, Jasmin A, Trask BJ, Hood L, Nelson PS (1998) Cloning and characterization of two Toll/Interleukin-1 receptor-like genes TIL3 and TIL4: evidence for a multi-gene receptor family in humans. Blood 91:4020–4027

Chesnokova O, Coutinho JB, Khan IH, Mikhail MS, Kado CI (1997) Characterization of flagella genes of *Agrobacterium tumefaciens*, and the effect of a bald strain on virulence. Mol Microbiol 23:579–590

Chilcott GS, Hughes KT (2000) Coupling of flagellar gene expression to flagellar assembly in *Salmonella enterica serovar typhimurium* and Escherichia coli. Microbiol Mol Biol Rev 64:694–708

Ciacci-Woolwine F, Blomfield IC, Richardson SH, Mizel SB (1998) Salmonella flagellin induces tumor necrosis factor alpha in a human promonocytic cell line. Infect Immun 66:1127–1134

Ciacci-Woolwine F, McDermott PF, Mizel SB (1999) Induction of cytokine synthesis by flagella from gram-negative bacteria may be dependent on the activation or differentiation state of human monocytes. Infect Immun 67:5176–5185

Dangl JL, Jones JD (2001) Plant pathogens and integrated defence responses to infection. Nature 411:826–833

Dibb-Fuller MP, Allen-Vercoe E, Thorns CJ, Woodward MJ (1999) Fimbriae- and flagella-mediated association with and invasion of cultured epithelial cells by *Salmonella enteritidis*. Microbiology 145:1023–1031

Dietrich C, Heuner K, Brand BC, Hacker J, Steinert M (2001) Flagellum of Legionella pneumophila positively affects the early phase of infection of eukaryotic host cells. Infect. Immun. 69:2116–2122

DiMango E, Zar HJ. Bryan R, Prince A (1995) Diverse *Pseudomonas aeruginosa* gene products stimulate respiratory epithelial cells to produce interleukin-8. J Clin Invest 96:2204–2210

Eaton KA, Suerbaum S, Josenhans C, Krakowka S (1996) Colonization of gnotobiotic piglets by *Helicobacter pylori* deficient in two flagellin genes. Infect Immun 64:2445–2448

Eaves-Pyles T, Murthy K, Liaudet L, Virag L, Ross G, Soriano FG, Szabo C, Salzman AL (2001) Flagellin, a novel mediator of *Salmonella*-induced epithelial activation and systemic inflammation: I κ B alpha degradation, induction of nitric oxide synthase, induction of proinflammatory mediators, and cardiovascular dysfunction. J Immunol 166:1248–1260

Eaves-Pyles T, Szabo C, Salzman AL (1999) Bacterial invasion is not required for activation of NF-κB in enterocytes. Infect Immun 67:800–804

Feldman M, Bryan R, Rajan S, Scheffler L, Brunnert S, Tang H, Prince A (1998) Role of flagella in pathogenesis of *Pseudomonas aeruginosa* pulmonary infection. Infect Immun 66:43–51

Felix G, Duran JD, Volko S, Boller T (1999) Plants have a sensitive perception system for the most conserved domain of bacterial flagellin. Plant J 18:265276

Fitzgerald KA, Palsson-McDermott EM, Bowie AG, Jefferies CA, Mansell AS, Brady G, Brint E, Dunne A, Gray P, Harte MT, et al (2001) Mal (MyD88-adapter-like) is required for Toll-like receptor-4 signal transduction. Nature 413:78–83

Fleiszig SM, Arora SK, Van R, Ramphal R (2001) FlhA, a component of the flagellum assembly apparatus of *Pseudomonas aeruginosa*, plays a role in internalization by corneal epithelial cells. Infect Immun 69:4931–4937

Gewirtz AT, Navas TA, Lyons S, Godowski PJ, Madara JL (2001a) Cutting edge: bacterial flagellin activates basolaterally expressed tlr5 to induce epithelial proinflammatory gene expression. J Immunol 167:1882–1885

Gewirtz AT, Siber AM, Madara JL, McCormick BA (1999) Orchestration of neutrophil movement by intestinal epithelial cells in response to *Salmonella typhimurium* can be uncoupled from bacterial internalization. Infect Immunology 67:608–617

Gewirtz AT, Simon PO, Jr., Schmitt CK, Taylor LJ, Hagedorn CH, O'Brien AD, Neish AS, Madara JL (2001b) *Salmonella typhimurium* translocates flagellin across intestinal epithelia, inducing a proinflammatory response. J Clin Invest 107:99–109

Girardin SE, Tournebize R, Mavris M, Page AL, Li X, Stark GR, Bertin J, DiStefano PS, Yaniv M, Sansonetti PJ, Philpott DJ (2001) CARD4/Nod1 mediates NF-kappaB and JNK activation by invasive *Shigella flexneri*. EMBO J. 2:736–742

Gomez-Gomez L, Bauer Z, Boller T (2001) Both the extracellular leucine-rich repeat domain and the kinase activity of FSL2 are required for flagellin binding and signaling in Arabidopsis. Plant Cell 13:1155–1163

Gomez-Gomez L, Boller T (2000) FLS2: an LRR receptor-like kinase involved in the perception of the bacterial elicitor flagellin in *Arabidopsis*. Mol Cell 5:1003–1011

Grant CC, Konkel ME, Cieplak W, Jr., Tompkins LS (1993) Role of flagella in adherence, internalization, and translocation of *Campylobacter jejuni* in nonpolarized and polarized epithelial cell cultures. Infect Immun 61:1764–1771

Hayashi F, Smith KD, Ozinsky A, Hawn TR, Yi EC, Goodlett DR, Eng JK, Akira S, Underhill DM, Aderem A (2001) The innate immune response to bacterial flagellin is mediated by Toll- like receptor 5. Nature 410:1099–1103

Hirano T, Yamaguchi S, Oosawa K, Aizawa S (1994) Roles of FliK and FlhB in determination of flagellar hook length in *Salmonella typhimurium*. J Bacteriol 176:5439–5449

Horng T, Barton GM, Medzhitov R (2001) TIRAP: an adapter molecule in the Toll signaling pathway. Nature Immunol 2:835–841

Humphrey T (2001) *Salmonella Typhimurium* definitive type 104. A multi-resistant Salmonella. Int J Food Microbiol 67:173–186

Ikeda JS, Schmitt CK, Darnell SC, Watson PR, Bispham J, Wallis TS, Weinstein DL, Metcalf ES, Adams P, O'Connor CD, O'Brien AD (2001) Flagellar phase variation of *Salmonella enterica serovar Typhimurium* contributes to virulence in the murine typhoid infection model but does not influence *Salmonella*-induced enteropathogenesis. Infect Immun 69:30213030

Imler JL, Hoffmann JA (2001) Toll receptors in innate immunity. Trends Cell Biol 11:304–311

Jarrossay D, Napolitani G, Colonna M, Sallusto F, Lanzavecchia A (2001) Specialization and complementarity in microbial molecule recognition by human myeloid and plasmacytoid dendritic cells. Eur J Immunol. 31:3388–3393

Kadowaki N, Ho S, Antonenko S, Malefyt RW, Kastelein RA, Bazan F, Liu YJ (2001) Subsets of human dendritic cell precursors express different Toll-like receptors and respond to different microbial antigens. J Exp Med 194:863–869

Kaisho T, Takeuchi O, Kawai T, Hoshino K, Akira S (2001) Endotoxin-induced maturation of MyD88-deficient dendritic cells. J Immunol 166:5688–5694

Kathariou S, Kanenaka R, Allen RD, Fok AK, Mizumoto C (1995) Repression of motility and flagellin production at 37 degrees C is stronger in *Listeria monocytogenes* than in the nonpathogenic species Listeria innocua. Can J Microbiol 41:572–577

Klose KE, Mekalanos JJ (1998) Differential regulation of multiple flagellins in *Vibrio cholerae*. J Bacteriol 180:303–316

Li X, Rasko DA, Lockatell CV, Johnson DE, Mobley HL (2001) Repression of bacterial motility by a novel fimbrial gene product. EMBO J 20:4854–4862

Liaudet L, Murthy KG, Mabley JG, Pacher P, Soriano FG, Salzman AL, Szabo C (2002) Comparison of Inflammation, Organ Damage, and Oxidant Stress Induced by *Salmonella enterica Serovar Muenchen* Flagellin and Serovar Enteritidis Lipopolysaccharide. Infect Immun 70:192–198

Lopez-Boado YS, Wilson CL, Parks WC (2001) Regulation of Matrilysin Expression in Airway Epithelial Cells by *Pseudomonas aeruginosa* Flagellin. J Biol Chem 276:41417–41423

McDermott PF, Ciacci-Woolwine F, Snipes JA, Mizel SB (2000) High-affinity interaction between gram-negative flagellin and a cell surface polypeptide results in human monocyte activation. Infect Immun 68:5525–5529

Medzhitov R (2001) Toll-like receptors and innate immunity. Nature Reviews Immunology 1:135–145

Medzhitov R, Janeway CA, Jr (1997) Innate immunity: the virtues of a nonclonal system of recognition. Cell 91:295–298

Meindl T, Boller T, Felix G (2000) The bacterial elicitor flagellin activates its receptor in tomato cells according to the address-message concept. Plant Cell 12:1783–1794

Michel T, Reichhart JM, Hoffmann JA, Royet J (2001) Drosophila Toll is activated by Gram-positive bacteria through a circulating peptidoglycan recognition protein. Nature 414:756–759

Mobley HL, Belas R, Lockatell V, Chippendale G, Trifillis AL, Johnson DE, Warren JW (1996) Construction of a flagellum-negative mutant of *Proteus mirabilis*: effect on internalization by human renal epithelial cells and virulence in a mouse model of ascending urinary tract infection. Infect Immun 64:5332–5340

Moors MA, Li L, Mizel SB (2001) Activation of interleukin-1 receptor-associated kinase by gram-negative flagellin. Infect Immun 69:4424–4429

Mostov KE, Verges M, Altschuler Y (2000) Membrane traffic in polarized epithelial cells. Curr Opin Cell Biol 12:483–490

Muzio M, Bosisio D, Polentarutti N, D'Amico G, Stoppacciaro A, Mancinelli R, van't Veer C, Penton-Rol G, Ruco LP, Allavena P, Mantovani A (2000) Differential expression and regulation of Toll-like receptors (TLR) in human leukocytes: selective expression of TLR3 in dendritic cells. J Immunol 164:5998–6004

Ogushi K, Wada A, Niidome T, Mori N, Oishi K, Nagatake T, Takahashi A, Asakura H, Makino S, Hojo H, et al (2001) *Salmonella enteritidis* FliC (flagella filament protein) induces human beta-defensin-2 mRNA production by Caco-2 cells. J Biol Chem 276:30521–30526

Peel M, Donachie W, Shaw A (1988) Temperature-dependent expression of flagella of *Listeria monocytogenes* studied by electron microscopy, SDS-PAGE and western blotting. J Gen Microbiol. 134:2171–2178

Rock FL, Hardiman G, Timans JC, Kastelein RA, Bazan JF (1998) A family of human receptors structurally related to *Drosophila* Toll. Proc Natl Acad Sci USA 95:588–593

Samakovlis C, Asling B, Boman HG., Gateff E, Hultmark D (1992) In vitro induction of cecropin genes – an immune response in a *Drosophila* blood cell line. Biochem Biophys Res Commun 188:1169–1175

Samatey FA, Imada K, Nagashima S, Vonderviszt F, Kumasaka T, Yamamoto M, Namba K (2001) Structure of the bacterial flagellar protofilament and implications for a switch for supercoiling. Nature 410:331–337

Schmitt CK, Ikeda JS, Darnell SC, Watson PR, Bispham J, Wallis TS, Weinstein DL, Metcalf ES, O'Brien AD (2001) Absence of all components of the flagellar export and synthesis machinery differentially alters virulence of *Salmonella enterica serovar Typhimurium* in models of typhoid fever, survival in macrophages, tissue culture invasiveness, and calf enterocolitis. Infect Immun 69: 5619–5625

Sebastiani G, Leveque G, Lariviere L, Laroche L, Skamene E, Gros P, and Malo D (2000). Cloning and characterization of the murine Toll-like receptor 5 (Tlr5) gene: sequence and mRNA expression studies in *Salmonella*-susceptible MOLF/Ei mice. Genomics 64:230–240

Sierro F, Dubois B, Coste A, Kaiserlian D, Kraehenbuhl JP, Sirard JC (2001) Flagellin stimulation of intestinal epithelial cells triggers CCL20- mediated migration of dendritic cells. Proc Natl Acad Sci USA 98:13722–13727

Sims JE (2002) IL-1 and IL-18 receptors, and their extended family. Curr Opin Immunol 14:117–122

Steiner TS, Nataro JP, Poteet-Smith CE, Smith JA, Guerrant RL (2000) Enteroaggregative *Escherichia coli* expresses a novel flagellin that causes IL-8 release from intestinal epithelial cells. J Clin Invest 105:1769–1777

Tauszig S, Jouanguy E, Hoffmann JA, Imler JL (2000) Toll-related receptors and the control of antimicrobial peptide expression in *Drosophila* Proc Natl Acad Sci USA 97:10520–10525

Tsang RS, Figueroa G, Bryden L, Ng L (2001) Flagella as a potential marker for *Campylobacter jejuni* strains associated with Guillain-Barre syndrome. J Clin Microbiol 39:762–764

Underhill DM, Ozinsky A. (2002) Toll-like receptors: key mediators of microbe detection. Curr Opin Immunol 14:103–110

Wyant TL, Tanner MK, Sztein MB (1999) *Salmonella typhi* flagella are potent inducers of proinflammatory cytokine secretion by human monocytes. Infect Immun 67:3619–3624

Zarember KA, Godowski PJ (2002) Tissue Expression of Human Toll-like receptors and differential regulation of Toll-like receptor mRNAs in leukocytes in response to microbes, their products, and cytokines. J Immunol 168:554–561

Zhang MY, Lovgren A, Landen R (1995) Adhesion and cytotoxicity of *Bacillus thuringiensis* to cultured *Spodoptera* and *Drosophila* cells. J Invertebr Pathol 66:46–51

TLR4 as the Mammalian Endotoxin Sensor

B. Beutler

For more than a century, the ability to sense endotoxin (later known also as lipopolysaccharide; LPS) stood as the archetypal innate immune response: even before the phrase 'innate immunity' became popular. Yet the mechanism by which LPS initiated a signal remained unknown. The problem was solved in 1998 by positional cloning, which revealed that Toll-like receptor (TLR) 4, one of ten mammalian paralogues with homology to the *Drosophila* protein Toll, is the central component of the LPS receptor. During the 3 years that followed, gene knockout work supported the view that the TLRs perceive a number of indispensable molecular structures shared by diverse representatives of the microbial world. The highly specific LPS-sensing function of TLR4 is remarkable for its prevalence in *Mammalia*, which to the present time is the only class of the phylum *Chordata* known to have a gene encoding TLR4, and known to display exquisite sensitivity to LPS. The fact that LPS signals are elicited through a single biochemical pathway has raised important pharmacotherapeutic opportunities as well.

1 The Roots of Innate Immunity, Planted in the Soil of Microbial Pathogenesis

The search for the *causa causans* – the 'cause of the cause' – is one that drives science in all disciplines. No sooner has a scientific conundrum been solved, than new questions present themselves. So it was that when Koch and Pasteur founded

Department of Immunology, The Scripps Research Institute, 10550 N. Torrey Pines Road, La Jolla, CA 92037, USA, *e-mail: brūce@scripps.edu*

the germ theory of disease, studies of pathogenesis entered an entirely new realm, confronting mysteries that had been inaccessible (and even undreamed of) before. The causal relationship between microbes and disease became widely accepted. But what was it about microbes that proved so injurious to the host? Rarely in a lethal infection was the microbial burden sufficient to explain the damage that was observed. Almost never were microbes present in blood in numbers sufficient to cause vaso-occlusion, for example, and in some cases, a lethal infection might not be associated with bacteremia at all. Obviously, poisons must emanate from the bacteria. The seeming 'toxicity' of an infection, known since antiquity, might at last be understood, provided that the toxic principles of infectious bacteria could be isolated and chemically characterized.

Among the first of the toxic principles to be identified, by L. Brieger and C. Fränkel in 1890, was an exotoxin (so-called toxalbumin) derived from filtrates of *Corynebacterium diphtheriae* cultures. It thus emerged that some of the poisons were soluble, exceptionally potent, and proteinaceous. But others were not proteins, were far more common and far more abundant than the rather exotic toxalbumin, and seemed to explain many of the characteristics that all infections shared with one another: fever, shock, and disturbed metabolism. Beginning with *Vibrio cholerae* organisms, R. Pfeiffer identified 'primary cholera toxin', a relatively insoluble, heat-stable material tightly associated with the bacterium itself (Pfeiffer 1892); hence, the subsequent use of the term 'endotoxin'.

It fell to later generations of microbiologists and chemists to learn that endotoxin was an abundant glycolipid component of Gram-negative bacteria (Boivin 1946; Boivin and Mesrobeanu 1935), and to deduce its chemical structure (Luderitz and Galanos 1973; Luderitz et al. 1966). For a still longer time, the fundamental question of how endotoxin 'worked' remained unsolved. Why, after all, was it toxic?

Over the years, endotoxin (ultimately known as lipopolysaccharide; LPS) came to be regarded as the archetypal microbial inducer of the septic syndrome; that is, as the molecule that most closely reproduced the effects of an authentic infection. But curiously, Gram-positive organisms (which do not produce LPS), and Gram-negative organisms, which do produce LPS, both create septic syndromes that are essentially indistinguishable in clinical terms. Hence, it was clear that other microbial factors must also have toxic effects, and it might have been guessed that all acted by triggering similar receptors. But because no other microbial toxin was so ubiquitous, so potent, or so readily isolated, LPS remained, until recent times, at the center of attention among workers concerned with microbial pathogenesis.

Among many informative curiosities in the LPS field was the fact that some species (e.g., humans, rabbits, and ungulates) were exquisitely sensitive to the toxin; others (rats, mice, and lower primates) were far less so. Extreme endotoxicity is a phenomenon known only in mammals. Birds, reptiles, amphibians, and fish are quite indifferent to LPS, at least in post-embryonic life. The haphazard representation of LPS toxicity among different branches of the phylogenic tree would suggest that a relatively small number of unique genes must make all the difference in determining LPS sensitivity.

States of LPS tolerance and hypersensitivity were described, and offered clues concerning pathogenesis, but only rather vague ones. Once challenged with LPS, an animal would be rendered refractory to subsequent LPS challenge for a period of hours to days. This phenomenon was seen to be reflected at the cellular level once it was realized that macrophages were among the principal cells capable of responding LPS in vivo. It was also observed that animals 'primed' by injection of various microbes such as *Corynebacterium parvum*, *Mycobacterim leprae*, and *M. bovis* would become sensitive to minute doses of LPS, insufficient to kill or injure non-sensitized controls. These observations hinted at a pathogenic mechanism for LPS, but did not define a clear objective at the protein level.

2 Many Modes of Toxicity, or Only One?

The complexity of physiologic derangements observed in the septic syndrome in mammals naturally creates the impression that a very complex mechanism must be at play. The fact that purified preparations of a single molecule – bacterial LPS – could recreate most of the syndrome was remarkable in itself. But more remarkable still was the realization that all of the untoward effects of LPS were channeled through a single host molecule.

While it was once imagined that LPS might act to perturb biological membranes, or bind to many receptors, its highly specific mode of action was clarified by a single, serendipitous discovery. In 1965, Heppner and Weiss observed that mice of the C3H/HeJ strain were highly resistant to the lethal effect of LPS. Subsequently, resistance was shown to be unique to the HeJ substrain, whereas all other C3H substrains (including C3H/HeN and C3H/OuJ, generally used as controls) were LPS sensitive. It was inferred, then, that a spontaneous mutation had become fixed in the C3H/HeJ population shortly after its creation, leading to total LPS insensitivity.

In due course, a single locus controlling LPS sensitivity was identified. This, the so-called *Lps* locus, was mapped to mouse chromosome 4 by Watson and colleagues in 1978 (Watson et al. 1977, 1978), who outcrossed mice of the C3H/HeJ strain to other animals bearing classical genetic markers. The mutation in C3H/HeJ mice, described as codominant because F1 animals showed an intermediate LPS response phenotype, was positioned between the *Mup-1* (major urinary protein) and *Ps* (polysndactyly) loci. In parallel with this, a second strain of *Mus musculus* (C57BL/10ScCr) was shown to be refractory to LPS (Coutinho et al. 1977), and through genetic analysis it was shown that the defect in these animals was allelic to the *Lps* defect of C3H/HeJ mice. Hence, F1 mice produced by crossing C57BL/10ScCr mice with C3H/HeJ mice were completely unresponsive to endotoxin, as were each of the parent strains (Coutinho and Meo 1978).

Although the molecular basis of endotoxin resistance caused by *Lps* mutations was at this point far from clear, it was well established that all biological effects of LPS depended upon the product of a single gene. Several hypotheses might have been offered to account for this. The most commonly held view was that that mice of the C3H/HeJ and C57BL/10ScCr strains had mutations in a unique receptor for LPS, which rendered them insensitive to its activating effect. If indeed *Lps* encoded a receptor, it would necessarily be a solitary receptor: multiple independent pathways for cell activation could not reasonably be contemplated. Alternatively, a single gene might be required to maintain 'endotoxin competence' in a number of other ways. Perhaps a unique transcription factor was required to respond to LPS; or perhaps a cytoplasmic protein kinase was lacking; indeed, no class of signaling protein could be excluded from consideration. But in all events, irrespective of the degree of ramification of the LPS signaling pathway within the cell, and irrespective of the number of protein contacts LPS might have outside the cell, all signals must converge upon a single protein at some point, and the notion of a single biochemical pathway for LPS sensing seemed secure.

Over the years, many studies performed with C3H/HeJ mice attested to the specificity of their genetic defect. Although taxol, a microtubule-binding drug derived from the Pacific Yew tree (*Taxus brevifolia*), was shown to have comparatively weak LPS-like activity (Ding et al. 1990a) [on which basis it was suggested that *Lps* might encode a microtubule-associated protein (Ding et al. 1990b)] no microbial product other than LPS has ever been convincingly shown to be incapable of activating C3H/HeJ macrophages, but capable of activating C3H/HeN macrophages. In time, 'endotoxin' came to have a biological definition matching its chemical identity as LPS: that which activates macrophages via the pathway defective in C3H/HeJ mice.

3 Concrete Ties Between Microbial Pathogenesis and Innate Immunity

'LPS the toxin' came to be seen as 'LPS the early-warning signal' with the demonstration that lymphoreticular cells were responsible for the untoward effects of endotoxin after its administration to animals. Whatever direct influence LPS might have on most tissues of the body, cells of hematopoietic origin were required for it to exert its lethal effect. So much was proven by Michalek et al. who used mice of the C3H/HeJ and C3H/HeN strains for that purpose (Michalek et al. 1980). By irradiating animals of each strain and reconstituting their hematopoietic population with progenitors from animals of the opposite strain, she and her colleagues demonstrated that the donor strain would determine the LPS susceptibility of the chimera.

Insofar as immune cells mediated the lethal effect of LPS, it might be guessed that LPS detection was strictly within the purview of the immune system, and also, that the untoward response of the host to LPS represented an exaggerated immune response. The picture was made still clearer by separate experiments of O'brien et al. who showed that C3H/HeJ mice were discretely immunocompromised (O'Brien et al. 1980). While normal animals could survive challenge with an intraperitoneal inoculum of 10^4 *Salmonella typhimurium* organisms, the C3H/HeJ mice were readily killed by as few as two organisms. In subsequent years, several other Gram-negative bacteria were also found to be more virulent in C3H/HeJ mice than in control strains (Cross et al. 1995; Macela et al. 1996; Mitchell et al. 1997; Yoshida et al. 1991). The most plausible interpretation was that LPS recognition permitted the rapid mobilization of an immune response: a response of the type that fell under the heading of innate immunity, as the entire experiment was of brief duration. Hence, LPS was not only a toxin, but also a target for innate immune detection.

3.1 Both Toxin and Target: Implications of the Dual Character of LPS

The undoubted structural function of LPS notwithstanding, the toxicity of LPS was once widely viewed as 'intentional,' from the standpoint of the bacterium. Indeed, some lipid A structures are far more toxic to a given mammalian species than others, and prior to the 1980 discovery that LPS sensing was essential for the effective clearance of a Gram-negative infection, it might reasonably have been argued that LPS was maintained as a 'weapon'; that Gram-negative bacteria evolved to achieve a level of virulence commensurate with their needs, based on the structure of the lipid A that they encoded.

However, a countervailing hypothesis now gained ascendancy. It appeared more likely that the host immune system had evolved to detect LPS, and had reserved a gene – the gene encoded at the *Lps* locus – for this express purpose. Among its virtues as a target, LPS could be seen as relatively conserved in structure. Moreover, it was represented in some form in nearly all Gram-negative species, and where it was present, it was essential for the integrity (or at least the fitness) of the Gram-negative organisms that expressed it.

The nature of innate immune receptors had always been elusive, and the fact that one type of receptor was devoted to LPS made it a particularly appealing receptor to find. Numerous efforts were made to do so. In some instances, LPS was used as a photoaffinity ligand; in other cases, affinity chromatography was attempted; in other cases, metabolically labeled LPS was used to identify the receptor. However, none of these approaches was successful, and in particular, none succeeded in establishing the genetic difference between C3H/HeJ and C3H/HeN mice. Likewise, attempts to find differences between the two strains at the protein level were fruitless. So, too, were expression cDNA cloning approaches. The *Lps* locus, it seemed, must encode a protein expressed at low copy number, perhaps bearing a very subtle change.

3.2 CD14 as a Concentrating Ligand for LPS

The nature of the LPS receptor began to emerge from studies peripheral to the *Lps* mutation itself. First, it was observed that a plasma protein, dubbed LBP, was capable of binding LPS in the plasma (TOBIAS et al. 1986). Produced by the liver, and acting as an acute phase reactant that was itself induced by LPS, LBP was then noted to opsonize LPS for recognition by a macrophage receptor (WRIGHT et al. 1989). In 1990, Wright et al. (WRIGHT et al. 1990) reported that CD14 was the receptor that recognized LPS–LBP complexes. However, although this sequence of events undoubtedly served to facilitate the detection of trace amounts of LPS present in plasma, it was insufficient to explain macrophage activation, as CD14, a glycosylphosphoinositol-tethered protein on the macrophage surface, was incapable of transducing a transmembrane signal.

Suspicion then grew that the 'missing' component of the LPS receptor complex must involve that protein encoded by the *Lps* locus. To the existing approaches already mentioned, directed attempts to find proteins that might complex with CD14 were then added. These attempts were unsuccessful, however, and in the end, a strict positional cloning approach was required to find the gene of interest.

4 Positional Cloning of *Lps* Reveals Identity to TLR4

Between 1993 and 1998, the *Lps* locus was mapped on 2093 meioses to an interval of mouse chromosome 4 that was 2.6 million base pairs in length. This interval was scoured for genes, using exon trapping, hybridization/selection, and DNA sequencing. Only one gene was identified in this immense interval, near the proximal border of the confinement. This gene encoded the murine version of Toll-like receptor (TLR)4 (POLTORAK et al. 1998a). DNA sequence analysis of the TLR4 cDNA from C3H/HeJ mice revealed a single base pair substitution, specifying the replacement of a conserved proline residue in the cytoplasmic domain of the receptor with a histidine (POLTORAK et al. 1998b). This substitution distinguished C3H/HeJ and C3H/HeN substrains, and to date it is the only mutational difference known to distinguish these substrains. In mice of the LPS-resistant C57BL/10ScCr strain, the *Tlr4* gene was not expressed (POLTORAK et al. 1998b). This proved to be due to a deletion spanning the entire locus (POLTORAK et al. 2000. The closely related, endotoxin-sensitive C57BL/10ScSn substrain had normal levels of TLR4 mRNA.

The fact that two independent, allelic LPS resistance mutations both involved *Tlr4* gave definitive proof that *Tlr4* was the essential gene in the *Lps* locus. The protein specified by *Tlr4* had the structural characteristics of a single-spanning transmembrane receptor. It seemed, then, that the long-sought co-receptor for CD14 had been found. The TLR4 protein, long studied as the product of *Lps*, had only one known function: the detection of endotoxin.

4.1 Structural and Functional Similarity to *Drosophila* Toll and to IL-1R

The discovery that *Lps* was identical to *Tlr4* was particularly important because of a phylogenetic connection not yet discussed in this chapter. No discrete function had been ascribed to any of the mammalian TLRs, five of which had been mined from expressed sequence tag databases at the time TLR4 was identified as the transducing component of the LPS receptor (Chaudhary et al. 1998; Medzhitov et al. 1997; Nomura et al. 1994; Rock et al. 1998; Taguchi et al. 1996). The TLRs were named for their structural similarity to *Drosophila* Toll, a single-spanning transmembrane receptor with leucine-rich repeats (LRR) in its ectodomain, and a Toll/IL-1 receptor (TIR) motif comprising most of its cytoplasmic domain.

The Toll protein of *D. melanogaster*, originally known for its developmental effects (Anderson et al. 1985), had been shown in 1996 to be required for the dipteran immune response to fungal infections (Lemaitre et al. 1996). Hence animals with deleterious mutations of Toll, its protein ligand Spätzle, or any of its downstream signaling partners were not capable of mounting an adequate response to the inoculation of *Aspergillus fumigatus*, or other fungi. One of nine *Drosophila* paralogues, Toll was clearly an innate immune receptor in flies. The identification of the *Lps* locus in mice made it clear that a mammalian homologue of Toll – TLR4 – also had a specific and well-defined innate immune sensing function.

Knockout mutations of other members of the TLR family were soon to show that at least several, and perhaps all of the mammalian TLRs, had been retained to serve the needs of innate immune detection (Alexopoulou et al. 2001; Hemmi et al. 2000; Takeguchi et al. 1999).

4.2 Molecular Details of Signal Transduction

While Toll interacts with a protein ligand and has no direct contact with microbial molecules of any sort, TLR4 has direct contact with LPS. The likelihood of interaction was first established by genetic complementation, in which cells of C3H/HeJ mice were transfected to express either normal human or normal mouse TLR4 protein. Tetra-acylated lipid A, an LPS partial structure that is known to activate mouse cells but not human cells, was capable of stimulating cells that expressed mouse TLR4 but not human TLR4. On the other hand, both cell lines responded to hexa-acylated lipid A (Poltorak et al. 2000).

Human TLR4 thus was able to make the 'decision' as to whether lipid A was tetra-acylated or hexa-acylated. Such a decision would almost certainly require close contact between the receptor and the agonist ligand. Similar results were reported by Lien et al., who made use of transfected Chinese hamster ovary cells rather than primary macrophages (Lien et al. 2000).

Independent analysis, using labeled LPS preparations, revealed contact between the ligand and CD14, TLR4, and MD-2, a small protein that is believed to be tightly associated with the TLR4 ectodomain (see below). Hence, both physical association studies and genetic studies lead to the conclusion that LPS has direct contact with TLR4. The fact of contact may be a general one, insofar as TLR9 appears to interact directly with unmethylated DNA, and like TLR4, shows a species-related preference for specific oligonucleotide ligands (BAUER et al. 2001).

4.3 The Identification of MD-2 as a Component of the LPS Receptor Complex

Numerous efforts have been made to establish in vitro, transfection-based systems for analysis of TLR signaling. In some instances, important errors of interpretation have been made. For example, TLR2 and CD14 were cotransfected into HEK 293 cells, which were subsequently stimulated with LPS. Nuclear factor (NF)-κB activation was then measured indirectly, by means of a luciferase-based reporter system (KIRSCHNING et al. 1998; YANG et al. 1998). The conclusion drawn from these experiments – that TLR2 is capable of transducing the LPS signal – was incorrect: an error which probably stemmed from impurities in the LPS preparations used, and the relative non-specificity of NF-κB signaling as an endpoint in such assays.

In the wake of the determination that LPS signals solely via TLR4, several groups attempted to reconstitute the LPS signaling pathway using this receptor rather than TLR2. Chow et al. reported immediate success (CHOW et al. 1999), in contrast to what had been reported by workers purporting to show signaling via TLR2 (YOSHIDA et al. 1991). On the other hand SHIMAZU et al. (SHIMAZU et al. 1999) maintained that no signal could be detected: absent cotransfection of an expression vector encoding MD-2, a small, secreted protein that appears to bind tightly to TLR4. Other workers (the author included) were never able to demonstrate LPS signal transduction in HEK293 cells under any conditions.

MD-2 was first identified as a homologue of MD-1 (MIURA et al. 1998), a protein that is tightly associated with the LRR-bearing ectodomain of rp105, a receptor that lacks a TIR but which otherwise resembles the TLRs (MIURA et al. 1996). The cotransfection study seemed to indicate that MD-2 was an essential component of the signaling apparatus. In further studies, the complex of MD-2 and TLR4 was also said to be required for taxol sensing (KAWASAKI et al. 2000).

The view that MD-2 is an essential part of the signaling complex is still open to challenge on the grounds that knockout mutations of MD-2 have not yet confirmed the need for this protein. A chemically induced mutation of MD-2 has been reported to be associated with LPS resistance in cultured Chinese hamster ovary cells (SCHROMM et al. 2001), and correction of the LPS response defect results from the expression of a normal MD-2 cDNA, introduced by transfection. On the surface, it seems peculiar that chemical mutagenesis could lead to a pure population of mRNA molecules bearing a mutation, as a heterozygous lesion would be expected.

However, it must be considered that the MD-2 locus might be functionally hemizygous in the cell line that was used. Granting this, genomic amplifications would be expected to show heterozygosity for the mutation; however, such data were not presented.

It is, of course, impossible to establish cause and effect in an in vitro model system using meiotic mapping. And transfection assays have proved misleading in the past, as noted earlier. It is to be hoped that a knockout mutation will soon provide a definitive answer concerning the role of MD-2.

5 The Evolution of LPS Sensing

As already mentioned, mammals are unique among vertebrates in their ability to sense LPS, and as described in Chap. 1 of this volume, the most recent gene duplication event that gave rise to TLR4 occurred at a time near to (or possibly before) the dawn of vertebrate evolution, approximately 500 million years ago. Because mammals and birds shared a common ancestor about 310 million years ago, it must be assumed that the precursor of modern TLR4 was present in both of these lines, and perhaps was once represented in other vertebrate lines as well. But most vertebrates have forsaken LPS sensing. Why, one might ask, is this? The calculations rendered by natural selection are difficult to understand, yet we do know that LPS sensing carries a substantial liability, and in some species, particularly some of those that live in an aqueous environment, this liability may have been intolerable.

Extreme sensitivity to LPS is witnessed in lower taxa. For example, in the horseshoe crab *Limulus polyphemus* in which LPS sensing is believed to be protective – as it is in mammals. Likewise, in the giant African snail, *Achatina fulica*, LPS sensing is known to occur (Biswas and Mandal 1999). In both cases, the lipid A moiety of LPS is sensed. In neither case is sensing known to be TLR-based, although a close relative of *Limulus polyphemus*, the Japanese horseshoe crab *Tachypleus tridentatus*, does have at least one TLR gene (Inamori et al. 2000). *Mollusca* and *Arthropoda* are distinct phyla, and there is no clear indication that a common mechanism applies, yet we have been surprised before.

It is likely that the need to sense LPS has arisen many times in the course of evolution, and on some occasions, a member of the TLR family has answered the call. On other occasions, a different solution has been found. An interesting question for the future arises from our newfound knowledge of conserved pathways among different species. Might it be advantageous to introduce TLR4 into species that lack it, in order to create LPS sensitivity? It might be imagined that domesticated birds, for example, often prey to Gram-negative infection, would benefit from an early detection pathway. But such is a suggestion that can only be addressed once the experiment has been undertaken.

6 Questions that Remain

It is likely that some components of the LPS receptor and signal transduction apparatus remain to be discovered. Genetic evidence attests to this. For example, a human proband with co-resistance to interleukin (IL)-1 and LPS has been identified (Kuhns et al. 1997), yet this individual has no obvious defect in the production of MyD88, MAL/TIRAP, or other proteins that serve LPS signaling.

Both biochemical approaches and genetic methods will be used to find all components of the LPS receptor, and it is an open question which type of approach is likely to bear fruit more quickly. Insofar as most genes have functions yet unknown, there is surely room for the discovery of quite a number of novel proteins that participate in signal transduction by LPS and the other members of this new family of receptors.

The reason for interspecies (and inter-individual) differences in LPS sensitivity remains unclear. So does the mechanism of LPS tolerance, and the mechanism of LPS hypersensitivity, or priming. However, there are at least logical points at which explanations may now be sought.

References

Alexopoulou L, Holt AC, Medzhitov R, Flavell RA (2001) Recognition of double-stranded RNA and activation of NF-κB by Toll- like receptor 3. Nature 413:732–8

Anderson KV, Jurgens G, Nusslein-Volhard C (1985) Establishment of dorsal-ventral polarity in the Drosophila embryo: genetic studies on the role of the Toll gene product. Cell 42:779–89

Bauer S, Kirschning CJ, Hacker H, Redecke V, Hausmann S, Akira S, Wagner H, Lipford GB (2001) Human TLR9 confers responsiveness to bacterial DNA via species-specific CpG motif recognition. Proc Natl Acad Sci USA 98:9237–42

Biswas C, Mandal C (1999) The role of amoebocytes in endotoxin-mediated coagulation in the innate immunity of Achatina fulica snails. Scand J Immunol 49:131–8

Boivin A (1946) Traveaux récents sur la constitution chimique et sur les propriétés biologiques des antigènes bactériens. Schweiz Z Pathol Bakteriol 9:505–41

Boivin A, Mesrobeanu L (1935) Recherches usr les antigènes somatiques et sur les endotoxines des bactéries. Rev Immunol 1:553–69

Chaudhary PM, Ferguson C, Nguyen V, Nguyen O, Massa HF, Eby M, Jasmin A, Trask BJ, Hood L, Nelson PS (1998) Cloning and characterization of two Toll/Interleukin-1 receptor-like genes TIL3 and TIL4: evidence for a multi-gene receptor family in humans. Blood 91:4020–7

Chow JC, Young DW, Golenbock DT, Christ WJ, Gusovsky F (1999) Toll-like receptor-4 mediates lipopolysaccharide-induced signal transduction. J Biol Chem 274:10689–92

Coutinho A, Forni L, Melchers F, Watanabe T (1977) Genetic defect in responsiveness to the B cell mitogen lipopolysaccharide. Eur J Immunol 7:325–8

Coutinho A, Meo T (1978) Genetic basis for unresponsiveness to lipopolysaccharide in C57BL/10Cr mice. Immunogenetics 7:17–24

Cross A, Asher L, Seguin M, Yuan L, Kelly N, Hammack C, Sadoff J, Gemski P Jr (1995) The importance of a lipopolysaccharide-initiated, cytokine- mediated host defense mechanism in mice against extraintestinally invasive *Escherichia coli*. J Clin Invest 96:676–86

Ding AH, Porteu F, Sanchez E, Nathan CF (1990) Downregulation of tumor necrosis factor receptors on macrophages and endothelial cells by microtubule depolymerizing agents. J Exp Med 171:715–27

Ding AH, Porteu F, Sanchez E, Nathan CF (1990) Shared actions of endotoxin and taxol on TNF receptors and TNF release. Science 248:370–2

Hemmi H, Takeuchi O, Kawai T, Kaisho T, Sato S, Sanjo H, Matsumoto M, Hoshino K, Wagner H, Takeda K, Akira S (2000) A Toll-like receptor recognizes bacterial DNA. Nature 408:740–5

Heppner G, Weiss DW (1965) High susceptibility of strain A mice to endotoxin and endotoxin-red blood cell mixtures. J Bacteriol 90:696–703

Inamori K, Koori K, Mishima C, Muta T, Kawabata S (2000) A horseshoe crab receptor structurally related to Drosophila Toll. J Endotoxin Res 6:397–9

Kawasaki K, Akashi S, Shimazu R, Yoshida T, Miyake K, Nishijima M (2000) Mouse toll-like receptor 4.MD-2 complex mediates lipopolysaccharide-mimetic signal transduction by Taxol. J Biol Chem 275:2251 –4

Kirschning CJ, Wesche H, Merrill AT, Rothe M (1998) Human toll-like receptor 2 confers responsiveness to bacterial lipopolysaccharide. J Exp Med 188:2091–7

Kuhns DB, Long Priel DA, Gallin JI (1997) Endotoxin and IL-1 hyporesponsiveness in a patient with recurrent bacterial infections. J Immunol 158:3959–64

Lemaitre B, Nicolas E, Michaut L, Reichhart JM, Hoffmann JA (1996) The dorsoventral regulatory gene cassette spatzle/Toll/cactus controls the potent antifungal response in Drosophila adults. Cell 86: 973–83

Lien E, Means TK, Heine H, Yoshimura A, Kusumoto S, Fukase K, Fenton MJ, Oikawa M, Qureshi N, Monks B, Finberg RW, Ingalls RR, Golenbock DT (2000) Toll-like receptor 4 imparts ligand-specific recognition of bacterial lipopolysaccharide. J Clin Invest 105:497–504

Luderitz O, Galanos C, Lehmann V, Nurminen M, Rietschel ET, Rosenfelder G, Simon M, Westphal O (1973) Lipid A: chemical structure and biological activity. J Infect Dis 128:29

Luderitz O, Staub AM, Westphal O (1966) Immunochemistry of O and R antigens of Salmonella and related Enterobacteriaceae. Bacteriol Rev 30:192–255

Macela A, Stulik J, Hernychova L, Kroca M, Krocova Z, Kovarova H (1996) The immune response against *Francisella tularensis* live vaccine strain in Lps^n and Lps^d mice. FEMS Immunol Med Microbiol 13:235–8

Medzhitov R, Preston-Hurlburt P, Janeway CA, Jr. (1997) A human homologue of the Drosophila Toll protein signals activation of adaptive immunity. Nature 388:394–7

Michalek SM, Moore RN, McGhee JR, Rosenstreich DL, Mergenhagen SE (1980) The primary role of lymphoreticular cells in the mediation of host responses to bacterial endotoxin. J Infec Dis 141:55–63

Mitchell CR, Kempton JB, Scott-Tyler B, Trune DR (1997) Otitis media incidence and impact on the auditory brain stem response in lipopolysaccharide-nonresponsive C3H/HeJ mice. Otolaryngol Head Neck Surg 117:459–64

Miura Y, Miyake K, Yamashita Y, Shimazu R, Copeland NG, Gilbert DJ, Jenkins NA, Inazawa J, Abe T, Kimoto M (1996) Molecular cloning of a human RP105 homologue and chromosomal localization of the mouse and human RP105 genes (Ly64 and LY64). Genomics 38:299–304

Miura Y, Shimazu R, Miyake K, Akashi S, Ogata H, Yamashita Y, Narisawa Y, Kimoto M (1998) RP105 is associated with MD-1 and transmits an activation signal in human B cells. Blood 92: 2815–22

Nomura N, Miyajima N, Sazuka T, Tanaka A, Kawarabayasi Y, Sato S, Nagase T, Seki N, Ishikawa K, Tabata S (1994) Prediction of the coding sequences of unidentified human genes. I. The coding sequences of 40 new genes (KIAA0001-KIAA0040) deduced by analysis of randomly sampled cDNA clones from human immature myeloid cell line KG-1. DNA Res 1:27–35

O'Brien AD, Rosenstreich DL, Scher I, Campbell GH, MacDermott RP, Formal SB (1980) Genetic control of susceptibility to Salmonella typhimurium in mice: role of the LPS gene. J Immunol 124:20–4

Pfeiffer R (1892) Untersuchungen über das Choleragift. Z Hygiene 11:393–412

Poltorak A, He X, Smirnova I, Liu M-Y, Van Huffel C, Du X, Birdwell D, Alejos E, Silva M, Galanos C, Freudenberg MA, Ricciardi-Castagnoli P, Layton B, Beutler B (1998) Defective LPS signaling in C3H/HeJ and C57BL/10ScCr mice: mutations in *Tlr4* gene. Science 282:2085–8

Poltorak A, Ricciardi-Castagnoli P, Citterio A, Beutler B (2000) Physical contact between LPS and Tlr4 revealed by genetic complementation. Proc Nat Acad Sci USA 97:2163–2167

Poltorak A, Smirnova I, Clisch R, Beutler B (2000) Limits of a deletion spanning *Tlr4* in C57BL/10ScCr mice. J Endotoxin Res 6:51–56

Poltorak A, Smirnova I, He XL, Liu MY, Van Huffel C, McNally O, Birdwell D, Alejos E, Silva M, Du X, Thompson P, Chan EKL, Ledesma J, Roe B, Clifton S, Vogel SN, Beutler B (1998) Genetic and physical mapping of the *Lps* locus- identification of the toll-4 receptor as a candidate gene in the critical region. Blood Cells Molecules & Diseases 24:340–55

Rock FL, Hardiman G, Timans JC, Kastelein RA, Bazan JF (1998) A family of human receptors structurally related to Drosophila Toll. Proc Natl Acad Sci USA 95:588–93

Schromm AB, Lien E, Henneke P, Chow JC, Yoshimura A, Heine H, Latz E, Monks BG, Schwartz DA, Miyake K, Golenbock DT (2001) Molecular genetic analysis of an endotoxin nonresponder mutant cell line: a point mutation in a conserved region of MD-2 abolishes endotoxin-induced signaling. J Exp Med 194:79–88

Shimazu R, Akashi S, Ogata H, Nagai Y, Fukudome K, Miyake K, Kimoto M (1999) MD-2, a molecule that confers lipopolysaccharide responsiveness on Toll-like receptor 4. J Exp Med 189:1777–1782

Taguchi T, Mitcham JL, Dower SK, Sims JE, Testa JR (1996) Chromosomal localization of TIL, a gene encoding a protein related to the Drosophila transmembrane receptor Toll, to human chromosome 4p14. Genomics 32:486–8

Takeuchi O, Hoshino K, Kawai T, Sanjo H, Takada H, Ogawa T, Takeda K, Akira S (1999) Differential roles of TLR2 and TLR4 in recognition of Gram-negative and Gram-positive bacterial cell wall components. Immunity 11:443–451

Tobias PS, Soldau K, Ulevitch RJ (1986) Isolation of a lipopolysaccharide-binding acute phase reactant from rabbit serum. J Exp Med 164:777–93

Watson J, Kelly K, Largen M, Taylor BA (1978) The genetic mapping of a defective LPS response gene in C3H/HeJ mice. J Immunol 120:422–4

Watson J, Riblet R, Taylor BA (1977) The response of recombinant inbred strains of mice to bacterial lipopolysaccharides. J Immunol 118:2088–93

Wright SD, Ramos RA, Tobias PS, Ulevitch RJ, Mathison JC (1990) CD14, a receptor for complexes of lipopolysaccharide (LPS) and LPS binding protein. Science 249:1431–3

Wright SD, Tobias PS, Ulevitch RJ, Ramos RA (1989) Lipopolysaccharide (LPS) binding protein opsonizes LPS-bearing particle for recognition by a novel receptor on macrophages. J Exp Med 170:1231–41

Yang R-B, Mark MR, Gray A, Huang A, Xie MH, Zhang M, Goddard A, Wood WI, Gurney AL, Godowski PJ (1998) Toll-like receptor-2 mediates lipopolysaccharide-induced cellular signaling. Nature 395:284–8

Yoshida S, Goto Y, Miyamoto H, Fujio H, Mizuguchi Y (1991) Association of Lps gene with natural resistance of mouse macrophages against Legionella pneumophila. FEMS Microbiol Immunol 4:51–6

TLR2: Cellular Sensor for Microbial and Endogenous Molecular Patterns

C.J. Kirschning[1] and R.R. Schumann[2]

Toll-like receptor (TLR) 2 is a member of the vertebrate protein family of TLRs that has been studied in substantial detail over the last years. The extracellular domain of the type I receptor molecule TLR2 contains 18 to 20 leucine rich repeat (LRR) and LRR like motives. The intracellular domain of TLR2 contains a Toll/IL-1 receptor/resistance protein typical TIR domain. After the first implication of TLR4 in immunity thereinafter followed by the discovery of the lipopolysaccharide signal transducer function of TLR4, TLR2 was the first of ten mammalian TLRs proven to be directly involved in recognition of pathogen associated molecular patterns (PAMPs). Among the TLR2 specific agonists are microbial products representing broad groups of species such as Gram-positive and Gram-negative bacteria, as well as mycobacteria, spirochetes, and mycoplasm. PAMP induced phagosomal localization of TLR2 and TLR2 dependent apoptosis have been shown. Complex formation with other molecules involved in pattern recognition such as CD14, MD2, TLR1, and TLR6 has been implicated for TLR2. Surprisingly even proteinaceous host material such as heat shock protein (HSP) 60 has been demonstrated to activate cells through TLR2. Thus, TLR2 may be a sensor and inductor of specific defense processes, including oxidative stress and cellular necrosis initially spurred by microbial compounds. Here we summarize the current knowledge on the structure and function of TLR2, which is far from being complete. Detailed understanding of the biology of TLR2 will probably contribute to the characterization of a number of infectious diseases and potentially help in the development of novel intervention strategies.

[1] Institute of Medical Microbiology, Immunology, and Hygiene, Technical University of Munich, 81675 Munich, Germany, *e-mail: carsten.kirschning@lrz.tum.de*
[2] Institute of Microbiology and Hygiene, Charité Medical Center, Humboldt-University, 10117 Berlin, Germany, *e-mail: ralf.schūmann@charite.de*

1 Introduction

Recognition of microorganisms by the host precedes the defense reaction. Inactivation or disintegration of a microbial cell does not necessarily abolish its potential to elicit immune responses; on the contrary cell wall compounds or molecules from within the microbial cell may be released and initiate further specific host reactions. Microbes are covered with and contain components commonly produced by broad groups of microbial species, so called pathogen associated molecular patterns (PAMPs). Immuno-stimulatory PAMPs are recognized through binding to soluble and cellular receptors expressed constitutively by host cells (MEDZHITOV and JANEWAY 1997). Examples of germline encoded pattern recognition receptors (PRRs) mediating innate immunity are mannose binding protein, lipopolysaccharide binding protein (LBP), scavenger receptors, and CD14 (FEARON and LOCKSLEY 1996; ADEREM and ULEVITCH 2000).

Somatic gene rearrangement followed by T and B cell selection and expansion results in generation of highly specific reactions mediating activation of a wide arsenal of effector mechanisms in vertebrates (AHMED and GRAY 1996). This process not only needs initialization but also requires several days to mount a fully active system. Adaptation of immunity or activation of readily acquired immunity are both triggered by innate immunity permanently prepared to go into immediate action and send appropriate signals such as antigen presentation and cytokine release, as well as to release effector molecules for direct host defense. While the degrees of specificity of innate and adaptive immune receptors differ, a common characteristic of both groups of receptors is their specificity for ligands from not necessarily endogenous, but principally exogenous, sources. This characteristic distinguishes these molecules from other cellular receptors such as cytokine receptors.

PAMPs are potential ligands of host receptors. A prototypic PAMP, lipopolysaccharide (LPS) is an integral compound of the outer membrane of Gram-negative bacteria and elicits, when injected, symptoms similar to those observed during acute bacterial infections (ULEVITCH and TOBIAS 1995; RIETSCHEL et al. 1996). LPS consists of lipid A anchoring the molecule to the outer bacterial cell membrane, as well as of a covalently linked polysaccharide chain. Its profound effects towards host cells and the thorough chemical analysis of its structure made LPS the prime agent for experimental elicitation and analysis of host responses towards bacterial infection. The first-line host receptors for bacterial LPS are LBP and CD14 (SCHUMANN et al. 1990; WRIGHT et al. 1990).

Many bacteria do not have a cell wall architecture containing a double membrane such as the one found in Gram-negative bacteria carrying LPS. Gram-positive bacteria and mycobacteria, for example, do not contain a double membrane or LPS, yet infection of the host with these microbes also leads to a profound immuno-stimulation. The crucial compounds in these bacteria have been identified and include peptidoglycan (PGN), lipoteichoic acid (LTA), glycoproteins, outer surface proteins, lipoproteins, and other proteins. A group of bacteria that is neither Gram-positive nor Gram-negative, but includes important pathogens such as Borrelia and Treponema are the spirochetes. The question of whether members of this family contain LPS has been controversial; however, they do contain other PAMPs such as glycolipids and proteins.

Besides exogenous PAMPs, host derived molecules for which the extracellular presence is restricted to pathogenic conditions are recognized by primary immune cells such as macrophages and dendritic cells (DCs). Such agonistic molecules might be formed from precursors under conditions of oxidative stress in serum, are released upon cellular necrosis, and the occurrence of some may depend on dietary conditions (Frantz et al. 2001; Lee et al. 2001; Li et al. 2001). Candidate proteins for endogenous PAMPs generally mediating endogenous signals have been identified. Members of the cellular chaperone protein family of heat shock proteins (HSPs) have been shown to potently activate immune responses. A stimulatory effect of HSP60 from endogenous, as well as exogenous sources, i.e., on immune cells, has been shown (Kol et al. 1998).

As has been evidenced over the last years, Toll-like receptors (TLRs) display specificity in recognizing PAMPs. TLR2 is the member of the TLR family with the largest number of different 'ligands' (or agonists) identified to date. However, it has been shown that TLR2 in particular multimerizes with receptor complexes determining specificity. Heterologous receptor interaction in innate immunity explains specificity to agonists sharing structural characteristics while differing in other aspects. This has been demonstrated for differently acylated bacterial lipopeptides (Takeuchi et al. 2001). Heterologous receptor complexes may be predetermined or assembled upon ligand interaction (Pfeiffer et al. 2001).

2 Discovery of TLRs – *Drosophila*, Vertebrates, and Signal Transduction

Screening of mutagenized *Drosophila* embryos for identification of proteins involved in dorsal–ventral polarization and patterning during fruit fly development yielded distinct mutants (Belvin and Anderson 1996). Characterization of a dominant mutation named *toll* led to identification of the membrane bound Toll receptor molecule (Belvin and Anderson 1996). An intracellular signaling pathway from Toll ligand Spaetzle to the transcription factor and Rel protein family member Dorsal, as well as further signaling pathways, were identified and elucidated to a large extent by genetic screens in *Drosophila* (Belvin and Anderson 1996; Ip and Davis 1998). Involvement of Toll signaling components in the anti-fungal response of the adult fruit fly, as well as of one out of eight *Drosophila* Toll homologues – 18 wheeler – in the anti-bacterial response has subsequently been demonstrated (Lemaitre et al. 1996; Williams et al. 1997; Silverman and Maniatis 2001).

Application of the sequences of the intracellular domain of the interleukin (IL)-1 receptor (R)-I and *Drosophila* Toll in database analyses resulted in identification of human Toll-like receptors. Randomly sequenced cDNA (RSC) 786/TIL/TLR1 and human Toll/TLR4 were the first two of the ten currently known human TLRs to be identified (MITCHAM et al. 1996; MEDZHITOV et al. 1997; CHUANG and ULEVITCH 2001). Induction of nuclear factor (NF)-κB through the TLR4 Toll/IL-1R/Resistance (TIR) domain proved conservation of the Toll signaling pathway in TLR function. TLR4-mediated signaling activated genes encoding proteins with roles in inflammatory processes such as IL-6 and costimulatory molecule B7.1 (MEDZHITOV et al. 1997). The abundance of TLR4 mRNA in the spleen and in peripheral blood lymphocytes (PBL) further suggested a role for TLR4 in immunity. Primary sequences, expression patterns, genomic localization, and functional data were presented for the first five human TLRs (ROCK et al. 1998; CHAUDHARY et al. 1998). Genomic analysis of LPS resistant mouse strains identified TLR4 as the *lps* gene product (POLTORAK et al. 1998; QURESHI et al. 1999). A TLR4 mutation that replaces proline by histidine at residue 712 in C3H/HeJ mice, as well as a TLR4 null allele in the C57BL/10ScCr strain were found by comparative sequencing of the *lps* locus (POLTORAK et al. 1998; QURESHI et al. 1999). LPS resistance of gene targeted TLR4$^{-/-}$ mice was demonstrated confirming the role of TLR4 as LPS signal transducer (HOSHINO et al. 1999).

Symptoms of inflammation are elicited by cellular release of inflammatory mediators such as tumor necrosis factor alpha (TNF-α) and IL-1β. Their receptors mediate signal transduction into the target cell resulting in specific cell activation (O'NEILL and DINARELLO 2000). Analysis of TNF type II receptor function led to the identification of the TNFR associated factor (TRAF) protein family, and TRAF6 was identified as a central IL-1 signal transducer (ROTHE et al. 1994; CAO et al. 1996). IL-1 receptor associated kinase (IRAK) and myeloid differentiation marker (MyD) 88 represent prototype IL-1 signaling molecules consisting of individual groups of proteins (MUZIO et al. 1997; WESCHE et al. 1997). Key downstream IL-1 signaling molecules include the transcription factor NF-κB, as well as the mitogen/stress activated protein kinases (M/SAPK) p38, c-Jun NH_2-terminal kinase (JNK), and extracellular signal-regulated kinases (ERK; p42/p44) 1/2. The kinase MEKK1 links the signal transduction chain from TRAF6 to the inhibitor of NF-κB kinase (IKK) α and IKK β (BAEUERLE and BALTIMORE 1996; SILVERMAN and MANIATIS 2001). Divergence of IL-1R/TLR induced JNK-signaling from NF-κB signaling upstream of TRAF6, as opposed to divergence of TNF-α induced NF-κB-signaling and JNK signaling at the level of TRAF2 was suggested (MUZIO et al. 1998; SONG et al. 1997). Analysis of TRAF6 deficient mice revealed divergence of IL-1R/TLR induced JNK signaling pathways and the NF-κB pathway downstream of TRAF6 (CHANG and KARIN 2001; LOMAGA et al. 1999).

The striking similarity between the intracellular domains of *Drosophila* Toll and vertebrate IL-1RI led to the definition of the TIR domain (O'NEILL and DINARELLO 2000). The TIR domain is defined by a consensus sequence motif shared by members of the IL-1- and the Toll/TLR receptor protein families including intracellular adapter molecules, as well as a group of plant resistance proteins. Based on functional characterization of Rel protein/NF-κB signaling pathways in *Drosophila* and vertebrates, functional and structural similarities within components involved in signaling from TIRs to Rel proteins/NF-κB have been identified (BAEUERLE and BALTIMORE 1996; BELVIN and ANDERSON 1996). The presence of a TIR domain determines the membership of its carrier protein in the IL-1R/TLR protein superfamily regardless of differences such as the presence of further distinct domains (O'NEILL and DINARELLO 2000).

3 TLR2 Structure

Like the prototype *Drosophila* Toll molecule vertebrate TLRs are type I transmembrane receptors. Their N-terminal extracellular domain is characterized by a sequence of leucine rich repeat (LRR) motifs flanked by a membrane proximal LRR C-terminal cysteine-rich domain (LRRCT) (as exemplified by TLR2, Figs. 1, 2). In contrast to *Drosophila* Toll, vertebrate TLRs do not contain a cysteine-rich domain within the LRR cluster of the extracellular domain. Reminiscent of Toll structure, the LRRCT is followed by a transmembrane domain and a C-terminal TIR domain encompassing 120–200 amino acid (aa) residues, respectively (O'NEILL and DINARELLO 2000).

The complete human and murine TLR2 cDNAs have been identified recently (HEINE et al. 1999; ROCK et al. 1998) (accession numbers U88878 and AF124741, respectively). The premature human TLR2 molecule is a polypeptide encompassing 785 aa residues, the largest number of which are leucine (15.4% of the total protein). The extracellular domain, the transmembrane domain, and the intracellular domain of TLR2 contain 13, 3, and 4 cysteine residues, respectively. The size of the mature protein is 95 kDa as shown by immunoblot analysis. The apparent difference from the calculated size of the unmodified polypeptide (approximately 10kDa) could be due to glycosylation or other post-translational modifications.

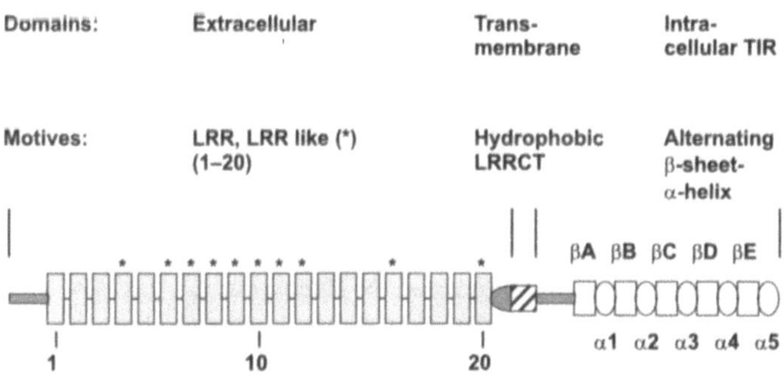

Fig. 1. TLR2 protein structure. Cartoon of the type I receptor molecule TLR2 extending from its extracellular amino terminus (*left*) to its intracellular C terminus (*right*). An N terminally located portion of TLR2 (*dark gray box*) is followed by 20 LRR and LRR-like motifs proposed (*light gray boxes*). Unmarked LRR motifs have been defined by application of the SMART program. LRR motifs marked with an *asterisk* have been defined based on their similarity to a consensus motif (see text and Fig. 3 for details). The transmembrane domain (*cross-hatched box*) is preceded by the membrane proximal LRR C-terminal motif (LRRCT; *dark gray half oval*) and is followed by a structurally undefined stretch (*second dark gray box*) and the TIR domain, respectively. The TIR domain forms a specific three-dimensional structure encompassing five alternating β-sheet (*open boxes*) and α-helix motifs (*open ovals*) pairs and provides the C terminus of the TLR2 molecule

MPHTLWMVWV LGVIISLSKE ESSNQASLSC DRNGICKGSS
GSLNSIPSGL

TEAVKSLDLSNN	RITYISNSDLQR	**LRR 1**
CVNLQALVLTSN	GINTIEEDSFSS	**LRR 2**
LGSLEHLDLSYN	YLSNLSSSWFKP	**LRR 3**
LSSLTFLNLLGN	PYKTLGETSLFSH	**LRR 4***
LTKLQILRVGNM	DTFTKIQRKDFAG	**LRR 5**
LTFLEELEIDAS	DLQSYEPKSLKS	**LRR 6***
IQNVSHLILHMK	QHILLLEIFVDV	**LRR 7***
TSSVECLELRDT	DLDTFHFSELSTGE	**LRR 8***
TNSLIKKFTFRN	VKITDESLFQVMKLLNQ	**LRR 9***
ISGLLELEFDDC	TLNGVGNFRASDNDRVID	**LRR 10***
PGKVETLTIRRL	HIPRFYLFYDLSTLYSL	**LRR 11***
TERVKRITVENS	KVFLVPCLLSQH	**LRR 12***
LKSLEYLDLSEN	LMVEEYLKNSACEDA	**LRR 13**
WPSLQTLILRQN	HLASLEKTGETLLT	**LRR 14**
LKNLTNIDISKN	SFHSMPETCQW	**LRR 15**
PEKMKYLNLSST	RIHSVTGCI	**LRR 16***
PKTLEILDVSNN	NLNLFSLN	**LRR 17**
LPQLKELYISRN	KLMTLPDASL	**LRR 18**
LPMLLVLKISRN	AITTFSKEQLDS	**LRR 19**
FHTLKTLEAGGN	*NFICSCEFLSFT*	**LRR 20***

QEQQALAK VLIDWPANYL CDSPSHVRGQ QVQDVRLSVS ECH
R
C-terminus of the extracellular domain of human
TLR2 (R587)

Fig. 2. Sequence of the extracellular domain of human TLR2 ordered by structural properties. Depicted is the amino acid sequence of the extracellular domain of human TLR2 (immature protein). Core motifs of the extracellular domain of TLR2 proposed hereby as LRR or LRR-like (*asterisk*) motifs are separated from the flanking sequences and aligned (*bold*). The motifs are numbered from 1 to 20; most LRR motifs, as well as an RNA binding domain (low significance, named LA, from L353 to T431) and a LRR C-terminal domain (italic) were identified by application of the SMART program (see text and Fig. 11). Amino acid residues principally conserved according to the minimal LRR consensus motif (LxxLxxLxLxxN; L, leucine; x, amino acid) are *underlined.* The C-terminally flanking two residues that are occupied by conserved leucines according to an extended LRR consensus motif (LxxLxxLxLxxNxLxxL) are *underlined* and fit the consensus sequence in three cases (LRR motif 3, 14, and 18). A potential translation start at amino acid M7 (*boxed*) results in calculation of a signal peptide encompassing the C-terminally flanking 20 amino acids

Several putative consensus sequences for post-translational modifications are present within human TLR2: it contains canonical N-glycosylation-, cAMP- and cGMP-dependent protein kinase phosporylation-, protein kinase C phosphorylation-, and N-myristoylation sites, and one aldehyde dehydrogenase glutamic acid active site within its extracellular domain as revealed by application of the SCANPROSITE program (http://www.expasy.ch/ tools/scripsit1.html/). Putative casein kinase II phosporylation sites are present within the extracellular and the intracellular domain of TLR2 as well.

TLR2 contains 18–20 LRR motifs within its extracellular domain as revealed by application of the SMART program (Schultz et al. 1998) for identification of motifs 1, 2, 3, and 5, as well as motifs 13–19 with the exception of motif 16 (http://smart.embl-heidelberg.de/) (Kobe and Deisenhofer 1994; Rock et al. 1998) (Fig. 2).

The other motifs were localized by definition of at least two matches within a LRR core motif (LxxLxxLxLxxN) as minimal requirement for a LRR-like motif (marked by asterisk in Figs. 1 and 2). The C-terminal region of the motif 20 may be part of the LRRCT thus disqualifying it as a LRR-like motif. LRR-like motif 12 lacks identity with the consensus sequence albeit displaying similarity (Fig. 2).

The TLR2 extracellular domain sequence displays particularly high similarity with the corresponding sequences of TLR1, TLR6, and TLR10 (Chuang and Ulevitch 2001). The similarity of TLR2 and the subgroup formed by TLR7, TLR8, and TLR9 is particularly low because of differences in the numbers and irregularities of LRRs, as well as non-LRR-like motif insertions. Comparison of the sequence spanning the region of LRR motifs numbers 13 to 16 in the extracellular domain of TLR2 (Fig. 2) and the membrane-proximal region of CD14 revealed considerable similarity implying functional resemblance of the two proteins (Kirschning and Bauer 2001). Application of the SMART sequence analysis program identifies a sequence from L352 to T431 (LLSQHLKSLE YLDLSENLMV EEYLKNSACE DAWPSLQTLI LRQNHLASLE KTGETLLTLK NLTNIDISKN SFHSMPET) based on its similarity to the RNA-binding domain of the Lupus La protein (Fig. 2). The motif spans from within LRR motif number 12 and extends to motif 15 of the extracellular domain of TLR2. Although some LRR-rich proteins are involved in DNA binding, identification of TLR9 and TLR3 as cellular signal mediators of microbial nucleic acids (Hemmi et al. 2000; Alexopoulou et al. 2001) lacking this motif speaks against this function of this domain. The crystal structure of one LRR protein, ribonuclease inhibitor, has been determined providing a model for other LRR proteins such as TLR stretches (Kajava 1998). Ribonuclease inhibitor LRR stretches represent units of one β-sheet and one α-helix, which are arranged in parallel to one axis. The β-sheets form an inner surface and the α-helices provide the interconnecting outer structure resulting in formation of a horseshoe-shaped molecule (Kajava 1998).

The transmembrane domain of TLR2 spans T588 to H610 of the premature protein. Tyrosine 617 located in close proximity, as well as the 'intra TIR' Y761, have been demonstrated to be phosphorylated upon activation of TLR2. Furthermore, they have been shown to be involved in NF-κB activation (Arbibe et al. 2000). Thus the N terminus of the intracellular domain of TLR2 not part of the TIR domain is actively involved in signal transduction via TLR2. The TIR domain starts 30 aa C-terminally of the transmembrane domain and extends to the C-terminal residue S785 of TLR2 (Rock et al. 1998; Xu et al. 2000). The intracellular TIR forms a cassette of ten motifs of alternating β-sheets and α-helices (Rock et al. 1998; Xu et al. 2000). Except for the interconnection of the third β-sheet and the third α-helix these ten alternating motifs are connected by eight loops. The structure of the loop between the second β-sheet and α-helix containing a sequence motif characteristic for IL-1R/TLRs has been a major focus of structure–function analysis (Xu et al. 2000). It includes amino acid residue P681 equivalent to mouse TLR4 P712 crucial for LPS responses (Poltorak et al. 1998; Qureshi et al. 1999). This so-called BB loop strongly contributes to a surface patch potentially important for interaction with the adapter molecule

MyD88 by forming a protrusion (Xu et al. 2000). Xu et al. proposed that although the structural changes caused by a P681H mutation in TLR2 are not significant, the residue P681 located at the tip of the loop may be important for interaction with other molecules such as those carrying TIR domains including MyD88 (Xu et al. 2000).

4 TLR2 Gene and Regulation

4.1 Genomic Organization and Promoter

The human TLR2 gene is localized at 4q32 in proximity to the TLR3 gene locus at 4q35. The TLR1 gene is localized on the same chromosome at 4p14, whereas the mouse TLR2 gene is located on chromosome 3 (Boyd et al. 2001; Rock et al. 1998). The three exons of the mouse gene span a genomic region of less than 10 kb in length. Exons one and two are located within the 5′-untranslated region of the TLR2 gene while the translation start codon is located15 base pairs downstream of the start codon of the third exon containing the entire coding sequence of mouse TLR2 (Musikacharoen et al. 2001; Wang et al. 2001). The sequence of the 5′-flanking region of the mouse TLR2 gene has been determined up to a position of 2.8 kb upstream. Transcriptional regulation of the TLR2 gene has been analyzed by reporter gene assays recently. Two NF-κB and two SP-1 recognition elements, as well as a signal transducer and activator of transcription (STAT) binding consensus sequence have been identified in the promoter sequence and found to be functionally involved in TLR2 gene regulation (Musikacharoen et al. 2001; Wang et al. 2001). Transcriptional gene regulation upon lipid A and IL-15 stimulation has been analyzed by reporter gene assays. A proximal NF-κB binding site governed IL-15 induced TLR2 gene activation in the mouse T cell line CTLL-2 substantially while in the mouse macrophage like cell line RAW264.7 this site was not involved in reporter gene activation (Musikacharoen et al. 2001). Constitutive expression of TLR2 apparently is maintained by recruitment of Sp1/Sp3 and NF-κB to the promoter. A minimal construct containing two NF-κB sites and one Sp1 recognition element conferred maximum transcriptional activation of the TLR2 gene in mouse monocytoid J774.1 cells upon infection with *Mycobacterium avium* as compared to further promoter constructs (Wang et al. 2001).

4.2 Regulation

Initial analysis of human TLR2 expression by Northern blotting revealed significantly increased TLR2 mRNA accumulation in the lung, spleen, and PBL as compared to other tissues (Chaudhary et al. 1998; Rock et al. 1998). Mouse TLR2 mRNA expression was found to be most pronounced in spleen, lung, and thymus (Matsuguchi et al. 2000). Treatment of primary mouse T cells from spleen and thymus with anti-CD3ε monoclonal antibody resulted in up-regulation of TLR2 mRNA accumulation as did stimulation of mouse T cell lines with IL-2, IL-15, or 4β-phorbol 12-myristate 13-acetate (PMA). PMA induced TLR2 expression occurred in an ERK1/2 and p38 dependent manner as revealed by application of specific kinase inhibitors (Matsuguchi et al. 2000). TLR2 was up-regulated in many organs including the liver upon an in vivo LPS challenge of mice. In mouse splenic macrophages TLR2 mRNA accumulation increased greatly upon stimulation with LPS or IL-15 as compared to mRNA levels in unstimulated cells. Stimulation of RAW264.7 mouse macrophage like cells with IL-2, IL-1β, interferon (IFN)-γ, as well as TNF-α also strongly induced TLR2 mRNA expression resulting in the proposal of a model in which up-regulation of TLR2 leads to sensitization of immune cells to TLR2 agonists (Matsuguchi et al. 2000). Infection of mouse macrophages with *Mycobacterium avium* resulted in a rapid increase of TLR2 mRNA levels in a TLR4-independent manner. In contrast, LPS induced TLR2 protein synthesis without inducing TLR2 mRNA up-regulation (Wang et al. 2001).

Human polymorph nuclear phagocytes (PMNs), as well as DCs, were shown to express TLR2 mRNA (Muzio et al. 2000). Upon LPS stimulation TLR2 mRNA levels were up-regulated in PMNs but not in monocytes. TLR2 mRNA expression furthermore was down-regulated during differentiation of monocytes to DCs (Muzio et al. 2000). Analysis of TLR2 protein expression in human cell lines was performed by several groups using flow cytometry. It was reported that the human myeloid cell lines THP1, U937, and Mono Mac 6 expressed TLR2 while other cell lines including the T lymphoblast Jurkat, the B lymphoblast Daudi, plasma cell RPMI 8226, microglioma U373, endothelial HMEC-1, epithelial HeLa, rhabdomsarcoma KYM-1, and fibroblast FS4 cells failed to express surface TLR2 (Flo et al. 2001). The $CD14^+$ subpopulation from whole blood cells expressed the highest levels of surface TLR2 (Flo et al. 2001). Monocyte-derived macrophages displayed TLR2 in the plasma membrane, in cytoplasmic compartments, and in the nucleus as revealed by confocal microscopy. IFN-γ has been shown to increase TLR2 surface staining in human peripheral blood mononuclear cells (Mita et al. 2001). LPS was shown to induce the rapid up-regulation of TLR2 mRNA levels in monocytes and immature DCs but not in mature TLR2 mRNA negative DCs (Visintin et al. 2001). Stimulation with LPS increased TLR2 mRNA accumulation also in endothelial cells and IFN-γ further enhanced this increase (Faure et al. 2001).

Application of *Aspergillus fumigatus* conidia also successively increased TLR2 mRNA accumulation in lungs of mice over a time period of 30 days, which could be of importance for mixed infections (Blease et al. 2001). Intra-peritoneal application of LPS to mice profoundly increased TLR2 mRNA levels in the central nervous system (CNS), while the application of PAMPs derived from Gram-positive bacteria failed to do so (Laflamme et al. 2001). TLR2 expression in primary intestinal epithelial cells was low as revealed by immunostaining and remained unchanged in samples from patients with ulcerative colitis or Crohn's disease. In the same study it was shown that patients with inflammatory bowel disease displayed significantly increased expression of TLR2 in the ileal or colonic epithelium (Cario and Podolsky 2000).

5 TLR2 Function

5.1 Recognition of Microbial Agonists

5.1.1 Recognition of PGN, LTA, Lipopeptides, LPS, and Additional Bacterial PAMPs

Gram-positive bacteria are of major clinical relevance and cause up to 50% of all cases of bacterial sepsis (Bone 1994). The main immuno-stimulatory cell wall components of Gram-positive bacteria are LTA and PGN (Heumann et al. 1994; De Kimpe et al. 1995). PGN is a constituent of virtually all bacteria but is present to a particularly high extent in the murein layer of the cell wall of Gram-positive bacteria. It consists of a polymeric glycan chain of alternating β (1,4) linked *N*-acetylmuramic acid and *N*-actylglucosaminyl monomers cross-linked by short peptide subunits (Schleifer and Kandler 1972). The peptides contain up to five amino acids and are linked to the N-acetylmuramic acid residues. LTA is a macroamphiphile present specifically in the cell wall of Gram-positive bacteria and consists of a poly-glycerophosphate backbone attached to a glycolipid (Fischer 2000). Like LPS, PGN and LTA elicit the release of pro-inflammatory cytokines such as TNF-α, IL-1β, and IL-6 from immune cells (Heumann et al. 1994).

Identification of human nucleotide sequences encoding *Drosophila* Toll-like proteins and determination of their cDNA sequences (Rock et al. 1998) were the basis for the initial implication of TLR2 in direct exogenous pattern recognition (Vogel 1998). The initial implication of TLR2 in recognition of Gram-positive bacteria was based on the application of whole bacteria, such as *Staphylococcus aureus*, or bacterial cell wall preparations to cells ectopically over-expressing TLR family members including TLR2 (Schwandner et al. 1999; Yoshimura et al. 1999). Application of highly purified soluble *S. aureus* PGN and commercially available *Bacillus subtilis* LTA identified both bacterial products as TLR2 agonists and further implied their role as PAMPs representing Gram-positive bacteria to the immune system (Schwandner et al. 1999; Yoshimura et al. 1999). While $TLR2^{-/-}$ mice displayed unresponsiveness to PGN, *S. aureus* LTA has been implicated as a TLR4 agonist (Takeuchi et al. 1999). In vitro analysis encompassing over-expression of MD2 in the human embryonic kidney fibroblast cell line HEK293 implicated both, TLR2 and TLR4, as LTA recognition molecules (Dziarski et al. 2001). Application of highly purified LTA preparations from various bacterial species such as *Treponema* and *S. aureus* to cell lines or primary mouse cells supported the earlier notion for TLR2 as an exquisite LTA signal transducer (Michelsen et al. 2001; Morath et al. 2001; Opitz et al. 2001).

Bacterial and mycoplasma lipopeptides are further TLR2 specific agonists. Modifications such as tripalmytilation and diacylation are typical (Bessler and Jung 1992; Muhlradt et al. 1997) and apparently important for this interaction. For a variety of different microbial lipopeptides including the soluble bacterial lipopeptide analogue synthetic tripalmitoyl-cysteinyl-seryl-(lysyl)$_3$-lysine (P3CSK4), OspA of *Borrelia burgdorferi*, and *Mycoplasma fermentans* macrophage activating lipopeptide (MALP)-2 a clear TLR2-dependent cell stimulation pattern has been observed (Aliprantis et al. 1999; Brightbill et al. 1999; Hirschfeld et al. 1999; Takeuchi et al. 2001). TLR2 specific cell activation by distinct bacterial lipopeptides has been demonstrated by application to human cell lines such as the glioma cell line U373, as well as by blocking of monocytic cell activation with a mAb (Aliprantis et al. 1999; Brightbill et al. 1999; Hirschfeld et al. 1999). Very low amounts of the lipopeptide MALP-2 from mycoplasma induced release of NO and TNF-α from mouse macrophages in a dose- and TLR2-dependent manner (Muhlradt et al. 1997; Takeuchi et al. 2001). Interestingly, the stereoisomeric orientation of this lipopeptide appears to be of major importance for the TLR2 interaction (Takeuchi et al. 2001). A cooperation of TLR2 with TLR6 or TLR1 in recognition of lipopeptides and other TLR2 agonists has been demonstrated (Ozinsky et al. 2000; Takeuchi et al. 2001).

LPS unresponsiveness despite CD14 over-expression paralleled by susceptibility to TNF-α and IL-1β qualified the HEK293 fibroblast cell line for complementation experiments with TLR protein family members. Responsiveness to commercially available LPS and lipid A preparations was gained by over-expression of human TLR2 in HEK293 cells (Yang et al. 1998; Kirschning et al. 1998). TLR2-dependent activation of NF-κB was specific as revealed by comparison with cellular effects of TLR1 and TLR4 over-expression. TLR1 failed to mediate cell activation while TLR4 over-expression caused constitutive activation of NF-κB (Muzio et al. 1998; Kirschning et al. 1998). Cell activation through TLR2 required the presence of the serum components LBP or soluble CD14 when low concentrations of LPS were applied (Yang et al. 1998; Kirschning et al. 1998). The implication of TLR4 as a LPS signal transducer by genetic analysis of mice (Poltorak et al. 1998; Qureshi et al. 1999) and also other findings challenged the implication of TLR2 in LPS signaling. Absence of functional TLR2 protein did not markedly impair LPS responsiveness of Chinese hamster ovary (CHO) cells (Heine et al. 1999). TLR2$^{-/-}$ mice were susceptible to shock by systemic application of high-dose LPS (Takeuchi et al. 1999). Special repurification of commercially available LPS preparations abrogated TLR2 but not TLR4-dependent cell activation in vitro (Hirschfeld et al. 2000). The phenotype of mice lacking functional TLR4 and their resistance towards LPS-induced shock clearly demonstrates the predominant role of TLR4 in response to LPS and thus Gram-negative bacteria in vivo. Although it is evident that TLR4 is the major LPS signal transducing receptor for classical LPS variants such as LPS from enterobacterial species, several lines of evidence support the implication of TLR2 as a PRR with gradually overlapping agonist specificity (Golenbock and Fenton 2001). As has been demonstrated for TLR4, MD2 might be a crucial component for TLR2-mediated pattern recognition as well and be required for LPS specificity (Akashi et al. 2001; Dziarski et al. 2001). Demonstration of LPS binding to TLR4 and TLR2 but not to other receptors of the TLR protein family, as well as impairment of the response of mouse TLR2$^{-/-}$ γδ T cells towards challenge with synthetic lipid A might indicate a role of TLR2 as a signaling LPS receptor relevant only under certain conditions in vivo (Mokuno et al. 2000; da Silva Correia et al. 2001). On the contrary, TLR4-independent use of TLR2 has been demonstrated for LPS from distinct bacterial species. LPS from *Porphyromonas gingivalis* has been shown to not only specifically recruit TLR2, but also to elicit partially different cellular responses as compared to *E. coli* LPS induced effects (Hirschfeld et al. 2001). Furthermore, LPS of the spirochete *Leptospira interrogans* has been demonstrated to require TLR2 for activation of immune cells in vitro and in vivo (Werts et al. 2001).

The first report of a TLR2 role in infection in vivo showed that gene targeted TLR2$^{-/-}$ mice displayed increased susceptibility to *S. aureus* infection as compared to wild-type mice (Takeuchi et al. 2000). Genetic analysis revealed a polymorphism in TLR2 that might be associated with increased susceptibility to *S. aureus* infection in humans (Lorenz et al. 2000). Group B streptococci (GBS), which are of major clinical importance for newborn children, display unique immuno-stimulatory activity; however, the molecule inducing immuno-stimulation has not yet been identified. It has been shown, however, that TLR2 plays a major role for recognition of GBS and the purification of the potentially novel TLR2 agonist is currently underway (Henneke et al. 2001). Recent reports indicate a function of TLR2 in recognition of bacterial fimbriae, although this function has previously been attributed to TLR4 (Asai et al. 2001; Frendeus et al. 2001). An unmodified synthetic peptide representing a fimbriae portion substituted for the whole protein in TLR2-mediated cell stimulation (Asai et al. 2001). A role of TLR2 in recognition of *Chlamydia pneumoniae*, which currently is discussed as being involved in the pathogenesis of arteriosclerosis, has been demonstrated (Prebeck et al. 2001). Finally, bacterial porins have been shown to use TLR2 for cell activation (Massari et al. 2002).

5.1.2 Spirochete PAMPs Recruit TLR2, as well as TLR2 and TLR4

TLR2 appears to be the major molecular sensor for Spirochetes, a class of bacteria characterized as neither Gram-positive nor Gram-negative (Porcella and Schwan 2001). These small, uniquely helical-shaped bacteria contain several immuno-stimulatory PAMPs and proved to be an interesting model for elucidating questions regarding structure–function relationship between TLRs and PAMPs. Spirochetes including the genera *Treponema* and *Borrelia,* as well as the family *Leptospira* are causative agents of a number of severe and frequently occurring chronic inflammatory diseases in humans, with Syphilis caused by *Treponema pallidum*, and Borreliosis or Lyme disease caused by *B. burgdorferi* being the most prominent ones and periodontitis being the most costly (Johnson 1977; Riviere et al. 1991). Impaired bacterial clearance of TLR2$^{-/-}$ mice in *B. burgdorferi* infection as compared to wild-type mice has been demonstrated (Wooten et al. 2002). In search for spirochetal compounds responsible for inflammatory reactions of the host, the presence of LPS in the outer membrane was reported repeatedly, however, the chemical analysis is so far incomplete. Analysis of the genome sequences of *T. pallidum* and *L. interrogans* yielded that *T. pallidum* lacks any genes involved in LPS synthesis with the exception of the *lic A* and *-C* genes. In *L. interrogans* in contrast, sequences displaying homology to genes involved in lipid A synthesis of Gram-negative bacteria have been found (Golenbock and Fenton 2001).

The search for biologically active products of these bacteria has focused on lipoproteins, although the presence of complex glycolipids distinct from LPS was never refuted. Some of these lipoproteins, referred to as outer surface proteins (Osps) A–F, were found to be strong inducers of pro-inflammatory cytokines in mononuclear cells via TLR2, as were spirochete lipoproteins (Hirschfeld et al. 1999). Two *Treponema* strains were recently analyzed chemically by using gas-liquid chromatography (GLC) and combined GLC-mass spectroscopy (GLC-MS) revealing the presence of glycolipids chemically distinct from LPS (Schroder et al. 2000). Further structural analysis led to the detection of a lipid anchor in the glycolipid resembling the structure of diacyl-glycerol lipid anchors present in LTA from Gram-positive bacteria. Experiments with two slightly different glycolipids from closely related *Treponema* strains using TLR4- and TLR2-deficient cells revealed that one glycolipid led to cytokine induction via TLR2, whereas the other utilized TLR2 and -4 (Schroder et al. 2001). These findings were supported by experiments using TLR-transfected HEK293 cells and over-expression of trans-dominant negative mutants of TLR–NF-κB signaling molecules such as MyD88 (Opitz et al. 2001).

5.1.3 Recognition of Mycobacteria, Protozoa, Fungi, and Others

TLR2 involvement in the mycobacterium–host-interaction has been confirmed by numerous studies. The prime focus on this interaction has been placed on lipoarabinomannan (LAM) of mycobacteria and this compound of the mycobacterial cell wall has been found to stimulate host cells via TLR2 (Means et al. 1999; Underhill et al. 1999). Interestingly, recognition of whole mycobacteria although involving TLR2 differs from LAM recognition (Means et al. 1999). Heat inactivation removed the activity of TLR4 agonists produced by *M. tuberculosis*, while TLR2-dependent PAMP recognition was not impaired relative to that of to live bacteria. Consequently, both quantitative and qualitative differences in the effect of viable and heat inactivated *M. tuberculosis* on gene activation profiles of DCs have been found (Ehrt et al. 2001). A recent report demonstrated a correlation between the frequency of leprosy caused by *M. leprae* and a TLR2 polymorphism, further implying a role of TLR2 in mycobacterial diseases (Kang and Chae 2001).

It has been shown that yeast zymosan induced cell activation but not necessarily uptake is TLR2 dependent (Underhill et al. 1999). Glycosylphosphatidylinositol anchors and glycoinositol phospholipids from parasitic protozoa have been shown to trigger NF-κB activation in CHO cells ectopically

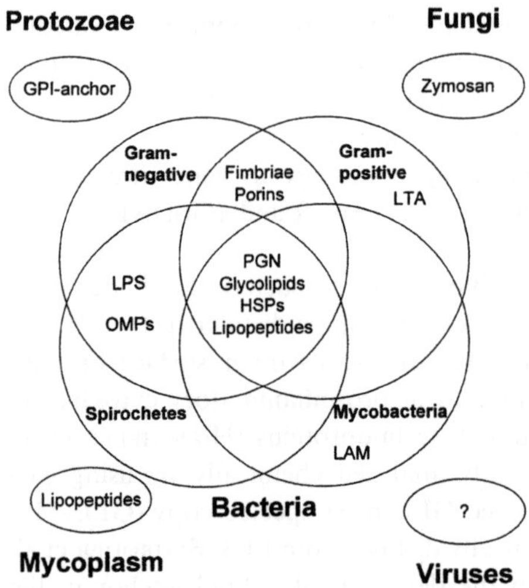

Fig. 3. Schematic overview for subgroups of bacterial species, mycoplasma, protozoae, and fungi, as well as for PAMPs recognized through TLR2. Gram-positive, Gram-negative, spirochetes, and mycobacteria as major bacterial groups, mycoplasma and eukaryotic microbes such as protozoa and fungi are depicted (*circles* or *ovals*). Some PAMPs are present in all bacterial species whereas others are present in subgroups only. Glycolipids encompass different subgroups of PAMPs such as mycobacterial lipoarbinomannans (*LAM*); zymosan is a lyophilized lysate of yeast cells from which a TLR2 agonist has not yet been identified. *OMPs*, Outer membrane proteins; *LPS*, lipopolysaccharide; *PGN*, soluble peptidoglycan; *LTA*, lipoteichoic acid; *HSPs*, bacterial heat shock proteins

over-expressing CD14 and TLR2, but not in wild-type CHO cells (CAMPOS et al. 2001). Analytical comparison of wild-type and TLR2 knockout mouse macrophages confirmed that TLR2 expression appears to be essential for induction of IL-12, TNF-α, and NO by glycosylphosphatidylinositol anchors derived from *Trypanosoma cruzi* trypomastigotes (CAMPOS et al. 2001). TLR4 has been excluded as signal transducer for a recombinant immuno-stimulatory protein of *Leishmania brasiliensis* implicating other TLRs as candidate receptors (BORGES et al. 2001). Figure 3 attempts to summarize some of the microbial PAMPs recognized through TLR2.

5.2 Recognition of Endogenous Agonists

The first report implicating TLR2 in recognition of endogenous, yet at that time unidentified, molecular patterns suggested a protective role against cell destruction mediated by TLR2. Oxidative stress impairs heart function via induction of cell death as well as hypertrophy of the myocardium. It was demonstrated that oxidative stress induced activation of p38 MAP kinase in primary rat myocytes (FRANTZ et al. 2001). This effect was not abrogated by application of an anti-TLR2 antiserum that did block activation of NF-κB and AP-1 otherwise induced by the same stimulus. The authors concluded that hydrogen peroxide induces activation of p38 but not of NF-κB and AP-1 in a TLR2-independent fashion. Thus TLR2-dependent cell activation may be brought about indirectly by prior induction of release of a TLR2 agonist. It was proposed by the authors that an agonist could be released from cells driven to necrosis or apoptosis by oxidative stress that may activate anti-apoptotic pathways protecting myocytes from death (FRANTZ et al. 2001).

A differentiation of TLR2-dependent pattern recognition between necrotic and apoptotic cells has been suggested by several groups. Lysed mouse embryonic fibroblasts (MEFs) and irradiated MEFs were applied as necrotic and apoptotic cells, respectively, for stimulation (Li et al. 2001). Cell preparations were applied to DCs, MEFs, macrophages, and transiently transfected HEK293 cells (Li et al. 2001). Apoptotic cells, either untreated or lysed, displayed less pronounced effects towards cells as compared to necrotic cells. Classical IL-1 receptor signal molecules such as MyD88 and TRAF6 were used in NF-κB activation as demonstrated by over-expression of dominant negative constructs in MEFs causing impairment of subsequent stimulation. Demonstration of TLR2-dependent cellular activation of HEK293 cells with necrotic cells complemented the study (Li et al. 2001).

Members of the heat shock protein family HSP60 of both human and bacterial origin were identified as potent stimuli of mouse macrophages in terms of TNF-α as well as of nitrogen monoxide (NO) release (Ohashi et al. 2000). Two lines of evidence suggested the involvement of TLR4 and TLR2 in cellular HSP60 recognition as proposed earlier: TNF-α release upon stimulation with chlamydial HSP60 of mouse immune cells of the C3H/HeJ strain lacking functional TLR4, as well as of $TLR2^{-/-}$ mouse DCs was profoundly decreased as compared to control cells (Ohashi et al. 2000; Vabulas et al. 2001). Secondly, over-expression of TLR4/MD2 or TLR2 in human epithelial cells was shown to confer responsiveness to both human and chlamydial HSP60 (Vabulas et al. 2001) see Chapter "Heat Shock Proteins as Ligands of Toll-Like Receptors" by Vabulas et al. in this book.

5.3 Cellular Activation, Phagocytosis, and Apoptosis

The recognition of molecular patterns includes signal transduction through the intracellular domains of cellular receptors such as TLRs. A main focus of IL-1R/TLR function analysis has been placed on the NF-κB and the S/MAPK inducing pathways (Lomaga et al. 1999; Silverman and Maniatis 2001) (Fig. 4). Biological parameters such as release of TNF-α, IL-1β, IL-6, and IL-12 have been analyzed for characterization of TLR functions. Also, induction of direct anti-microbial activity through TLR2 such as release of NO has been demonstrated (Thoma-Uszynski et al. 2001) (Fig. 4). In addition, subcellular translocation of TLR2 from the cell surface to phagosomes within 5 min upon microbial challenge has been shown (Underhill et al. 1999). Over-expression of a dominant-negative (dn) TLR2 point- or a dn MyD88 deletion mutant (TLR2-P681H or MyD88-C-terminus) did inhibit TNF-α synthesis but not internalization of TLR2 specific microbial particles in a murine macrophage cell line. TLR2 thus had been suggested to mediate cytokine response immediately upon phagocytosis of specific PAMPs (Underhill et al. 1999).

Whether TLR2 activates distinct signaling pathways as compared to those triggered by other TLR family members and whether immune responses towards particular pathogens differ has been addressed widely recently. Application of lipid A or LPS variants from two different bacterial species, *E. coli* and

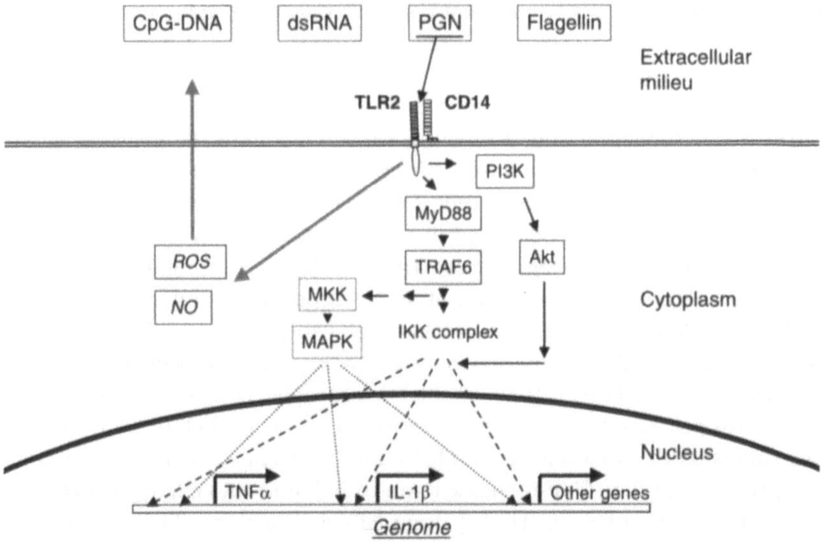

Fig. 4. TLR2 is a cellular detector for distinct molecular patterns. TLR2 and CD14 are expressed and presented on first-line immune cells that constitutively sample the extracellular milieu (extracellular space or phagosomal content) for molecular patterns that signal the need for immune responses (e.g., in the case of infection). Illustrated is cellular activation by soluble peptidoglycan (*PGN*) through TLR2 in the presence of further PAMPs (DNA, bacterial unmethylated CpG DNA; RNA, double-stranded viral RNA; flagellin, bacterial flagella protein). Homo or hetero di- or oligomerization may be required for TLR2 activation but is not depicted. Cell activation leads to synthesis and release of nitrogen monoxide (*NO*) and reactive oxygen species (*ROS*; *italic, boxed*; *dark gray arrows* for signal also depending on gene activation and release), as well as of inflammatory mediators such as proinflammatory cytokines TNF-α, IL-1β, and others. *Black arrows*, pattern recognition encompassing signal transduction via mediating molecules. IKKs and an unknown kinase downstream of Akt trigger nuclear translocation of and gene activation through NF-κB (*broken arrows*), while MAPK activates transcription factors such as AP-1 (*dotted arrows*) from cytoplasm to genes. Gene activation upon binding of transcription factors to recognition elements in regulatory regions culminates in synthesis and release of proteins involved in host response; transcription start sides are depicted as *angled arrows*. CD14 and MD2, as well as TLR1 and TLR6 might form large complexes containing further receptor chains and TLR2 for pattern recognition. *PI3K*, phospahtidylinositol 3 kinase; *Akt*, cellular protein kinase/PKB; *MyD88*, myeloid differentiation marker 88; *TRAF6*, TNF receptor associated factor 6; *IKK complex*, a multi-protein complex in which the inhibitor of NF-κB kinase (*IKK*) α and -β phosphorylate the inhibitor of NF-κB (I-κB) leading to release of NF-κB; *MAPK*, mitogen activated kinases (representing JNK, p38, and erk1/2); *MKK*, the specific kinase phosphorylating MAPK (for example, MKK3/6 for p38)

Porphyromonas gingivalis, differing in structural characteristics such as the number of fatty acids, revealed striking differences in the degree, time course, and quality of gene induction (Hirschfeld et al. 2001; Martin et al. 2001). These differences included distinct levels of IL-12p40 mRNA. mRNA accumulation was increased greatly upon stimulation of mouse macrophages with *E. coli* lipid A while being absent upon application of *P. gingivalis* lipid A (Hirschfeld et al. 2001). Levels of proinflammatory cytokines such as TNF-α released from human myeloid cells were markedly lower upon stimulation with *P. gingivalis* LPS as compared to *E. coli* LPS (Martin et al. 2001). While *E. coli* LPS was a stronger inducer of T helper

(Th)1 cytokines IFN-γ and IL-12, *P. gingivalis* LPS preferentially induced Th2 cytokines such as IL-10 in immune cells (PULENDRAN et al. 2001). *P. gingivalis* LPS did not induce tolerance to a second challenge with the same LPS (MARTIN et al. 2001). TLR2 over-expression experiments, as well as application of inhibiting antibodies identified *P. gingivalis* lipid A as a TLR2 agonist (HIRSCHFELD et al. 2001; MARTIN et al. 2001). Mediation of cross-tolerance by application of TLR2 and TLR4 agonists in mouse macrophages was demonstrated thus indicating the capacity of TLR2 to confer tolerance (SATO et al. 2000). Whether different TLR2 ligands elicit distinct effects or whether different affinities cause activation to specific degrees remains to be analyzed.

Bifurcation of the IKKα/β and the p38-mediated pathways at the level of the transforming growth factor β activated kinase (TAK) 1 has been postulated (SHUTO et al. 2001). Degradation of IκB as well as phosphorylation of p38 were impaired in epithelial cells over-expressing mutant dominant-negative TAK1 protein upon stimulation with the Gram-negative bacterium non-typeable *Haemophilus influenzae* (NTHi) as compared to mock transfected cells. Application of a blocking anti TLR2 antibody inhibited NTHi- and major NTHi outer membrane protein P6-mediated cell activation. It was proposed that p38 is involved in NF-κB p65 phosphorylation via yet unknown signal mediators (SHUTO et al. 2001). Besides IKK-dependent IκB degradation leading to release of NF-κB and translocation into the nucleus, p65 phosphorylation via TLR2 upon recruitment of the p85 subunit of phosphatidylinositol-3 kinase (PI3 K) to the phosphorylated TLR2 intracellular domain has been shown (ARBIBE et al. 2000). While involvement of the kinase Akt downstream from PI3 K has been demonstrated, the NF-κB p65 kinase itself remained elusive (ARBIBE et al. 2000) (Fig. 4). Transcription factors such as NF-κB and AP-1, as well as early growth response factor (Egr)1, cAMP response element binding factor CREB, and serum response element binding factor (SRE)-1 have been implicated in TLR-mediated gene activation (XU et al. 2001). The degrees and time kinetics of LPS and PGN-induced activation of transcription factors have been demonstrated to be similar (XU et al. 2001). TLR2 not only mediated cellular activation but also apoptotic effects. The TLR2-dependent pathways of NF-κB activation and apoptosis diverged at the level of MyD88 as revealed from over-expression of dominant negative constructs of MyD88 and TRAF6. Fas associated death domain protein and caspase 8 were implicated in a TLR2-dependent apoptotic pathway (ALIPRANTIS et al. 2000).

6 Perspectives

Reminiscent of the lack of impairment of mouse morphogenesis in any NF-κB subunit knockout mouse analyzed, and in spite of Dorsal function in *Drosophila* embryogenesis, no morphogenic impairment of any Toll-like receptor knockout mouse has been reported. Differences seem to extend further to characteristics such as the mechanism of pattern recognition through Tolls in *Drosophila* and TLRs in vertebrates. While Toll is activated upon binding of a proteolytically activated endogenous ligand named Spaetzle, a variety of indications speak for direct interaction of the TLR2 extracellular domain with its agonists thus qualifying them as ligands. Although strong evidence for direct agonist interaction has been presented for TLR4, analysis of TLR–agonist interaction is lacking for the other TLR protein family members.

Binding of LPS to TLR2 has been demonstrated (DA SILVA CORREIA et al. 2001). PGN, as well as MD2 and surfactant protein A have also been shown to bind to TLR2, leading to enhancement or inhibition of function, respectively (AKASHI et al. 2001; DZIARSKI et al. 2001; MURAKAMI et al. 2001). This could be interpreted as Spaetzle-equivalent or anti-Spaetzle functions of the proteins. Four mammalian peptidoglycan recognition proteins (PGRPs), one of them soluble, have been identified and characterized (LIU et al. 2001). One PGRP of *Drosophila*, PGRP-SA, has been demonstrated to be specifically involved in Toll activation upon challenge by Gram-positive bacteria such as *Streptococcus faecalis* and *Bacillus megaterium* (MICHEL et al. 2001). The identification of a family of PGRPs in humans suggests that the perspective on vertebrate PAMP recognition might have to be reconsidered. As under serum-free conditions PGN induced cell activation via TLR2 is enhanced as compared to that in serum-containing medium (SCHWANDNER et al. 1999), a proposed PRR and protease, as well as a pre-Späetzle-like ligand might be cell membrane bound such as exemplified by some members of the vertebrate protein family of PGRPs (LIU et al. 2001) (Fig. 5B). On the other hand, results from analysis of species-specific PAMP recognition further support the 'TLR = PRR' model (BAUER et al. 2001) (Fig. 5A).

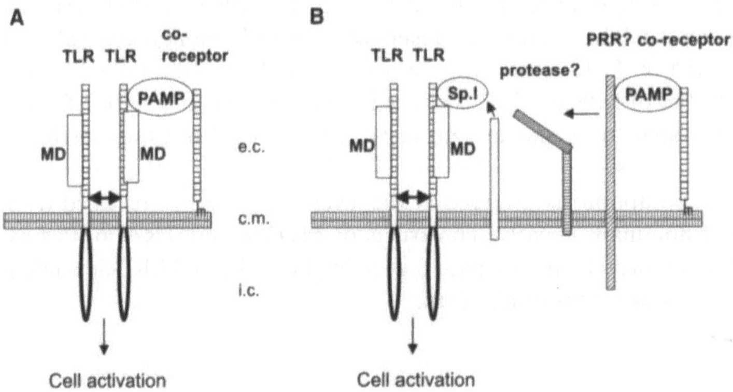

Fig. 5A,B. Models for PAMP recognition through TLRs. Two different models for cellular recognition of agonists via TLRs are proposed. According to the first model (**A**), TLR is a PRR that binds in cooperation with coreceptor chains such as CD14 and members of the MD protein family members (PAMPs) leading to TLR dimerization and cell activation. (**B**) Reminiscent of Toll function in *Drosophila* TLR may not be the PRR (*PRR?*) itself but may receive an indirect signal via a membrane bound protease cascade (*protease?*, see text) resulting in activation of a Spaetzle-like ligand (*Sp.l*) through precursor cleavage. Alternatively, the PRR and the protease function may be merged in a single molecule. *MD*, member of the MD receptor protein family such as MD-2; *e.c.*, extra cellular space; *c.m.*, cellular membrane; *i.c.*, intracellular space

Major and detailed insights in to the understanding of TLR2-mediated induction of mediator or effector synthesis have been gained by using classical methods to analyze the gene induction that results in expression of cytokines, as well as effector molecules such as anti-bacterial peptides (Birchler et al. 2001; Hirschfeld et al. 2001). More holistic perspectives of the changes in the host cell's gene activation pattern upon microbial challenge have been gained by micro array based gene profiling as compared to Northern blot analysis. As TLR2 agonistic PAMPs have been applied at least in the context of whole bacteria in some of these studies, the contribution of TLR2-dependent effects to the gene activation patterns observed can be implied. In one study expression patterns of mouse macrophages upon challenge with live or heat killed *M. tuberculosis* and/or IFN-γ have been analyzed by micro-array assays (Ehrt et al. 2001). Although overlapping, the spectra of genes induced by the two stimuli differ significantly (Ehrt et al. 2001). From the relevant biological species groups (*or*: groups of species) known to contain members producing immuno-stimulatory molecular patterns, i.e. viruses and plants have not been implicated in TLR2-mediated cell activation to date. Considering the striking diversity of TLR2 agonists it is tempting to speculate that they might produce molecular patterns employing TLR2 for cell activation as well. Further analysis of the contribution of TLR2 to the recognition of molecular patterns that induce immune responses in host organisms will enable further evaluation of its potential as therapeutic target for immune modulation under clinical conditions such as acute or chronic inflammation.

Acknowledgements. We thank T. Miethke, N. Schröder, and G. Lipford for helpful comments.

References

Aderem A, Ulevitch RJ (2000) Toll-like receptors in the induction of the innate immune response. Nature 406:782–7

Ahmed R, Gray D (1996) Immunological memory and protective immunity: understanding their relation. Science 272:54–60

Akashi S, Nagai Y, Ogata H, Oikawa M, Fukase K, Kusumoto S, Kawasaki K, Nishijima M, Hayashi S, Kimoto M, Miyake K (2001) Human MD-2 confers on mouse Toll-like receptor 4 species-specific lipopolysaccharide recognition. Int Immunol 13:1595–9

Alexopoulou L, Holt AC, Medzhitov R, Flavell RA (2001) Recognition of double-stranded RNA and activation of NF-κB by Toll-like receptor 3. Nature 413:732–8

Aliprantis AO, Yang RB, Mark MR, Suggett S, Devaux B, Radolf JD, Klimpel GR, Godowski P, Zychlinsky A (1999) Cell activation and apoptosis by bacterial lipoproteins through toll-like receptor-2. Science 285:736–9

Aliprantis AO, Yang RB, Weiss DS, Godowski P, Zychlinsky A (2000) The apoptotic signaling pathway activated by Toll-like receptor-2. Embo J 19:3325–36

Arbibe L, Mira JP, Teusch N, Kline L, Guha M, Mackman N, Godowski PJ, Ulevitch RJ, Knaus UG (2000) Toll-like receptor 2-mediated NF-κ B activation requires a Rac1-dependent pathway. Nat Immunol 1:533–40

Asai Y, Ohyama Y, Gen K, Ogawa T (2001) Bacterial fimbriae and their peptides activate human gingival epithelial cells through Toll-like receptor 2. Infect Immun 69:7387–95

Baeuerle PA, Baltimore D (1996) NF-κ B: ten years after. Cell 87:13–20

Bauer S, Kirschning CJ, Hacker H, Redecke V, Hausmann S, Akira S, Wagner H, Lipford GB (2001) Human TLR9 confers responsiveness to bacterial DNA via species-specific CpG motif recognition. Proc Natl Acad Sci U S A 98:9237–42
Belvin MP, Anderson KV (1996) A conserved signaling pathway: the Drosophila toll-dorsal pathway. Annu Rev Cell Dev Biol 12:393–416
Bessler WG, Jung G (1992) Synthetic lipopeptides as novel adjuvants. Res Immunol 143:548–53; discussion 579–80
Birchler T, Seibl R, Buchner K, Loeliger S, Seger R, Hossle JP, Aguzzi A, Lauener RP (2001) Human Toll-like receptor 2 mediates induction of the antimicrobial peptide human beta-defensin 2 in response to bacterial lipoprotein. Eur J Immunol 31:3131–7
Blease K, Kunkel SL, Hogaboam CM (2001) IL-18 modulates chronic fungal asthma in a murine model; putative involvement of Toll-like receptor-2. Inflamm Res 50:552–60
Bone RC (1994) Gram-positive organisms and sepsis. Arch Intern Med 154:26–34
Borges MM, Campos-Neto A, Sleath P, Grabstein KH, Morrissey PJ, Skeiky YA, Reed SG (2001) Potent stimulation of the innate immune system by a Leishmania brasiliensis recombinant protein. Infect Immun 69:5270–7
Boyd Y, Goodchild M, Morroll S, Bumstead N (2001) Mapping of the chicken and mouse genes for toll-like receptor 2 (TLR2) to an evolutionarily conserved chromosomal segment. Immunogenetics 52:294–8
Brightbill HD, Libraty DH, Krutzik SR, Yang RB, Belisle JT, Bleharski JR, Maitland M, Norgard MV, Plevy SE, Smale ST, Brennan PJ, Bloom BR, Godowski PJ, Modlin RL (1999) Host defense mechanisms triggered by microbial lipoproteins through toll-like receptors. Science 285:732–6
Campos MA, Almeida IC, Takeuchi O, Akira S, Valente EP, Procopio DO, Travassos LR, Smith JA, Golenbock DT, Gazzinelli RT (2001) Activation of Toll-like receptor-2 by glycosylphosphatidylinositol anchors from a protozoan parasite. J Immunol 167:416–23
Cao Z, Xiong J, Takeuchi M, Kurama T, Goeddel DV (1996) TRAF6 is a signal transducer for interleukin-1. Nature 383:443–6
Cario E, Podolsky DK (2000) Differential alteration in intestinal epithelial cell expression of toll-like receptor 3 (TLR3) and TLR4 in inflammatory bowel disease. Infect Immun 68:7010–7
Chang L, Karin M (2001) Mammalian MAP kinase signalling cascades. Nature 410:37–40
Chaudhary PM, Ferguson C, Nguyen V, Nguyen O, Massa HF, Eby M, Jasmin A, Trask BJ, Hood L, Nelson PS (1998) Cloning and characterization of two Toll/Interleukin-1 receptor-like genes TIL3 and TIL4: evidence for a multi-gene receptor family in humans. Blood 91:4020–7
Chuang T, Ulevitch RJ (2001) Identification of hTLR10: a novel human Toll-like receptor preferentially expressed in immune cells. Biochim Biophys Acta 1518:157–61
da Silva Correia J, Soldau K, Christen U, Tobias PS, Ulevitch RJ (2001) Lipopolysaccharide is in close proximity to each of the proteins in its membrane receptor complex. transfer from CD14 to TLR4 and MD-2. J Biol Chem 276:21129–35
De Kimpe SJ, Kengatharan M, Thiemermann C, Vane JR (1995) The cell wall components peptidoglycan and lipoteichoic acid from Staphylococcus aureus act in synergy to cause shock and multiple organ failure. Proc Natl Acad Sci USA 92:10359–63
Dziarski R, Wang Q, Miyake K, Kirschning CJ, Gupta D (2001) MD-2 enables Toll-like receptor 2 (TLR2)-mediated responses to lipopolysaccharide and enhances TLR2-mediated responses to Gram-positive and Gram-negative bacteria and their cell wall components. J Immunol 166:1938–44
Ehrt S, Schnappinger D, Bekiranov S, Drenkow J, Shi S, Gingeras TR, Gaasterland T, Schoolnik G, Nathan C (2001) Reprogramming of the macrophage transcriptome in response to interferon-gamma and Mycobacterium tuberculosis: signaling roles of nitric oxide synthase-2 and phagocyte oxidase. J Exp Med 194:1123–40
Faure E, Thomas L, Xu H, Medvedev A, Equils O, Arditi M (2001) Bacterial lipopolysaccharide and IFN-gamma induce Toll-like receptor 2 and Toll-like receptor 4 expression in human endothelial cells: role of NF-kappa B activation. J Immunol 166:2018–24
Fearon DT, Locksley RM (1996) The instructive role of innate immunity in the acquired immune response. Science 272:50–3
Fischer W (2000) Phosphocholine of pneumococcal teichoic acids: role in bacterial physiology and pneumococcal infection. Res Microbiol 151:421–7
Flo TH, Halaas O, Torp S, Ryan L, Lien E, Dybdahl B, Sundan A, Espevik T (2001) Differential expression of Toll-like receptor 2 in human cells. J Leukoc Biol 69:474–81
Frantz S, Kelly RA, Bourcier T (2001) Role of TLR-2 in the activation of nuclear factor κB by oxidative stress in cardiac myocytes. J Biol Chem 276:5197–203

Frendeus B, Wachtler C, Hedlund M, Fischer H, Samuelsson P, Svensson M, Svanborg C (2001) Escherichia coli P fimbriae utilize the Toll-like receptor 4 pathway for cell activation. Mol Microbiol 40:37–51

Golenbock DT, Fenton MJ (2001) Extolling the diversity of bacterial endotoxins. Nat Immunol 2:286–8

Heine H, Kirschning CJ, Lien E, Monks BG, Rothe M, Golenbock DT (1999) Cutting edge: cells that carry A null allele for toll-like receptor 2 are capable of responding to endotoxin. J Immunol 162:6971–5

Hemmi H, Takeuchi O, Kawai T, Kaisho T, Sato S, Sanjo H, Matsumoto M, Hoshino K, Wagner H, Takeda K, Akira S (2000) A Toll-like receptor recognizes bacterial DNA. Nature 408:740–5

Henneke P, Takeuchi O, van Strijp JA, Guttormsen HK, Smith JA, Schromm AB, Espevik TA, Akira S, Nizet V, Kasper DL, Golenbock DT (2001) Novel engagement of CD14 and multiple toll-like receptors by group B streptococci. J Immunol 167:7069–76

Heumann D, Barras C, Severin A, Glauser MP, Tomasz A (1994) Gram-positive cell walls stimulate synthesis of tumor necrosis factor alpha and interleukin-6 by human monocytes. Infect Immun 62:2715–21

Hirschfeld M, Kirschning CJ, Schwandner R, Wesche H, Weis JH, Wooten RM, Weis JJ (1999) Cutting edge: inflammatory signaling by Borrelia burgdorferi lipoproteins is mediated by toll-like receptor 2. J Immunol 163:2382–6

Hirschfeld M, Ma Y, Weis JH, Vogel SN, Weis JJ (2000) Cutting edge: repurification of lipopolysaccharide eliminates signaling through both human and murine toll-like receptor 2. J Immunol 165: 618–22

Hirschfeld M, Weis JJ, Toshchakov V, Salkowski CA, Cody MJ, Ward DC, Qureshi N, Michalek SM, Vogel SN (2001) Signaling by toll-like receptor 2 and 4 agonists results in differential gene expression in murine macrophages. Infect Immun 69:1477–82

Hoshino K, Takeuchi O, Kawai T, Sanjo H, Ogawa T, Takeda Y, Takeda K, Akira S (1999) Cutting edge: Toll-like receptor 4 (TLR4)-deficient mice are hyporesponsive to lipopolysaccharide: evidence for TLR4 as the Lps gene product. J Immunol 162:3749–52

Ip YT, Davis RJ (1998) Signal transduction by the c-Jun N-terminal kinase (JNK) from inflammation to development. Curr Opin Cell Biol 10:205–19

Johnson RC (1977) The spirochetes. Annu Rev Microbiol 31:89–106

Kajava AV (1998) Structural diversity of leucine-rich repeat proteins. J Mol Biol 277:519–27

Kang TJ, Chae GT (2001) Detection of Toll-like receptor 2 (TLR2) mutation in the lepromatous leprosy patients. FEMS Immunol Med Microbiol 31:53–8

Kirschning CJ, Bauer S (2001) Toll-like receptors: cellular signal transducers for exogenous molecular patterns causing immune responses. Int J Med Microbiol 291:251–60

Kirschning CJ, Wesche H, Merrill Ayres T, Rothe M (1998) Human toll-like receptor 2 confers responsiveness to bacterial lipopolysaccharide. J Exp Med 188:2091–7

Kobe B, Deisenhofer J (1994) The leucine-rich repeat: a versatile binding motif. Trends Biochem Sci 19:415–21

Kol A, Sukhova GK, Lichtman AH, Libby P (1998) Chlamydial heat shock protein 60 localizes in human atheroma and regulates macrophage tumor necrosis factor-alpha and matrix metalloproteinase expression. Circulation 98:300–7

Laflamme N, Soucy G, Rivest S (2001) Circulating cell wall components derived from gram-negative, not gram-positive, bacteria cause a profound induction of the gene-encoding Toll-like receptor 2 in the CNS. J Neurochem 79:648–57

Lee JY, Sohn KH, Rhee SH, Hwang D (2001) Saturated fatty acids, but not unsaturated fatty acids, induce the expression of cyclooxygenase-2 mediated through Toll-like receptor 4. J Biol Chem 276:16683–9

Lemaitre B, Nicolas E, Michaut L, Reichhart JM, Hoffmann JA (1996) The dorsoventral regulatory gene cassette spatzle/Toll/cactus controls the potent antifungal response in *Drosophila* adults. Cell 86: 973–83

Li M, Carpio DF, Zheng Y, Bruzzo P, Singh V, Ouaaz F, Medzhitov RM, Beg AA (2001) An essential role of the NF-κB/Toll-like receptor pathway in induction of inflammatory and tissue-repair gene expression by necrotic cells. J Immunol 166:7128–35

Liu C, Xu Z, Gupta D, Dziarski R (2001) Peptidoglycan recognition proteins: a novel family of four human innate immunity pattern recognition molecules. J Biol Chem 276:34686–94

Lomaga MA, Yeh WC, Sarosi I, Duncan GS, Furlonger C, Ho A, Morony S, Capparelli C, Van G, Kaufman S, van der Heiden A, Itie A, Wakeham A, Khoo W, Sasaki T, Cao Z, Penninger JM, Paige

CJ, Lacey DL, Dunstan CR, Boyle WJ, Goeddel DV, Mak TW (1999) TRAF6 deficiency results in osteopetrosis and defective interleukin-1, CD40, and LPS signaling. Genes Dev 13:1015–24

Lorenz E, Mira JP, Cornish KL, Arbour NC, Schwartz DA (2000) A novel polymorphism in the toll-like receptor 2 gene and its potential association with staphylococcal infection. Infect Immun 68:6398–401

Martin M, Katz J, Vogel SN, Michalek SM (2001) Differential induction of endotoxin tolerance by lipopolysaccharides derived from Porphyromonas gingivalis and Escherichia coli. J Immunol 167:5278–85

Massari P, Henneke P, Ho Y, Latz E, Golenbock DT, Wetzler LM (2002) Cutting Edge: Immune Stimulation by Neisserial Porins Is Toll-Like Receptor 2 and MyD88 Dependent. J Immunol 168:1533–1537

Matsuguchi T, Musikacharoen T, Ogawa T, Yoshikai Y (2000a) Gene expressions of Toll-like receptor 2, but not Toll-like receptor 4, is induced by LPS and inflammatory cytokines in mouse macrophages. J Immunol 165:5767–72

Matsuguchi T, Takagi K, Musikacharoen T, Yoshikai Y (2000b) Gene expressions of lipopolysaccharide receptors, toll-like receptors 2 and 4, are differently regulated in mouse T lymphocytes. Blood 95:1378–85

Means TK, Wang S, Lien E, Yoshimura A, Golenbock DT, Fenton MJ (1999) Human toll-like receptors mediate cellular activation by Mycobacterium tuberculosis. J Immunol 163:3920–7

Medzhitov R, Janeway CA (1997) Innate immunity: the virtues of a nonclonal system of recognition. Cell 91:295–8

Medzhitov R, Preston-Hurlburt P, Janeway CA, Jr (1997) A human homologue of the Drosophila Toll protein signals activation of adaptive immunity. Nature 388:394–7

Michel T, Reichhart JM, Hoffmann JA, Royet J (2001) Drosophila Toll is activated by Gram-positive bacteria through a circulating peptidoglycan recognition protein. Nature 414:756–9

Michelsen KS, Aicher A, Mohaupt M, Hartung T, Dimmeler S, Kirschning CJ, Schumann RR (2001) The role of toll-like receptors (TLRs) in bacteria-induced maturation of murine dendritic cells (DCS). Peptidoglycan and lipoteichoic acid are inducers of DC maturation and require TLR2. J Biol Chem 276:25680–6

Mita Y, Dobashi K, Shimizu Y, Nakazawa T, Mori M (2001) Toll-like receptor 2 and 4 surface expressions on human monocytes are modulated by interferon-gamma and macrophage colony-stimulating factor. Immunol Lett 78:97–101

Mitcham JL, Parnet P, Bonnert TP, Garka KE, Gerhart MJ, Slack JL, Gayle MA, Dower SK, Sims JE (1996) T1/ST2 signaling establishes it as a member of an expanding interleukin-1 receptor family. J Biol Chem 271:5777–83

Mokuno Y, Matsuguchi T, Takano M, Nishimura H, Washizu J, Ogawa T, Takeuchi O, Akira S, Nimura Y, Yoshikai Y (2000) Expression of toll-like receptor 2 on gamma delta T cells bearing invariant V gamma 6/V delta 1 induced by Escherichia coli infection in mice. J Immunol 165:931–40

Morath S, Geyer A, Hartung T (2001) Structure-Function Relationship of Cytokine Induction by Lipoteichoic Acid from Staphylococcus aureus. J Exp Med 193:393–8

Muhlradt PF, Kiess M, Meyer H, Sussmuth R, Jung G (1997) Isolation, structure elucidation, and synthesis of a macrophage stimulatory lipopeptide from Mycoplasma fermentans acting at picomolar concentration. J Exp Med 185:1951–8

Murakami S, Iwaki D, Mitsuzawa H, Sano H, Takahashi H, Voelker DR, Akino T, Kuroki Y (2001) Surfactant protein A inhibits peptidoglycan-induced TNF-alpha secretion in U937 cells and alveolar macrophages by direct interaction with toll-like receptor 2. J Biol Chem 27:27

Musikacharoen T, Matsuguchi T, Kikuchi T, Yoshikai Y (2001) NF-κB and STAT5 play important roles in the regulation of mouse Toll-like receptor 2 gene expression. J Immunol 166:4516–24

Muzio M, Bosisio D, Polentarutti N, D'Amico G, Stoppacciaro A, Mancinelli R, van't Veer C, Penton-Rol G, Ruco LP, Allavena P, Mantovani A (2000) Differential expression and regulation of toll-like receptors (TLR) in human leukocytes: selective expression of TLR3 in dendritic cells. J Immunol 164:5998–6004

Muzio M, Natoli G, Saccani S, Levrero M, Mantovani A (1998) The human toll signaling pathway: divergence of nuclear factor κB and JNK/SAPK activation upstream of tumor necrosis factor receptor-associated factor 6 (TRAF6). J Exp Med 187:2097–101

Muzio M, Ni J, Feng P, Dixit VM (1997) IRAK (Pelle) family member IRAK-2 and MyD88 as proximal mediators of IL-1 signaling. Science 278:1612–5

O'Neill LA, Dinarello CA (2000) The IL-1 receptor/toll-like receptor superfamily: crucial receptors for inflammation and host defense. Immunol Today 21:206–9

Ohashi K, Burkart V, Flohe S, Kolb H (2000) Cutting edge: heat shock protein 60 is a putative endogenous ligand of the toll-like receptor-4 complex. J Immunol 164:558–61
Opitz B, Schroder NW, Spreitzer I, Michelsen KS, Kirschning CJ, Hallatschek W, Zahringer U, Hartung T, Gobel UB, Schumann RR (2001) Toll-like receptor-2 mediates Treponema glycolipid and lipoteichoic acid-induced NF-κB translocation. J Biol Chem 276:22041–7
Ozinsky A, Underhill DM, Fontenot JD, Hajjar AM, Smith KD, Wilson CB, Schroeder L, Aderem A (2000) The repertoire for pattern recognition of pathogens by the innate immune system is defined by cooperation between toll-like receptors. Proc Natl Acad Sci USA 97:13766–71
Pfeiffer A, Bottcher A, Orso E, Kapinsky M, Nagy P, Bodnar A, Spreitzer I, Liebisch G, Drobnik W, Gempel K, Horn M, Holmer S, Hartung T, Multhoff G, Schutz G, Schindler H, Ulmer AJ, Heine H, Stelter F, Schutt C, Rothe G, Szollosi J, Damjanovich S, Schmitz G (2001) Lipopolysaccharide and ceramide docking to CD14 provokes ligand-specific receptor clustering in rafts. Eur J Immunol 31:3153–64
Poltorak A, He X, Smirnova I, Liu MY, Huffel CV, Du X, Birdwell D, Alejos E, Silva M, Galanos C, Freudenberg M, Ricciardi-Castagnoli P, Layton B, Beutler B (1998) Defective LPS signaling in C3H/HeJ and C57BL/10ScCr mice: mutations in Tlr4 gene. Science 282:2085–8
Porcella SF, Schwan TG (2001) Borrelia burgdorferi and Treponema pallidum: a comparison of functional genomics, environmental adaptations, and pathogenic mechanisms. J Clin Invest 107:651–6
Prebeck S, Kirschning C, Durr S, da Costa C, Donath B, Brand K, Redecke V, Wagner H, Miethke T (2001) Predominant role of toll-like receptor 2 versus 4 in Chlamydia pneumoniae-induced activation of dendritic cells. J Immunol 167:3316–23
Pulendran B, Kumar P, Cutler CW, Mohamadzadeh M, Van Dyke T, Banchereau J (2001) Lipopolysaccharides from distinct pathogens induce different classes of immune responses in vivo. J Immunol 167:5067–76
Qureshi ST, Lariviere L, Leveque G, Clermont S, Moore KJ, Gros P, Malo D (1999) Endotoxin-tolerant mice have mutations in Toll-like receptor 4 (Tlr4). J Exp Med 189:615–25
Rietschel ET, Brade H, Holst O, Brade L, Muller-Loennies S, Mamat U, Zahringer U, Beckmann F, Seydel U, Brandenburg K, Ulmer AJ, Mattern T, Heine H, Schletter J, Loppnow H, Schonbeck U, Flad HD, Hauschildt S, Schade UF, Di Padova F, Kusumoto S, Schumann RR (1996) Bacterial endotoxin: Chemical constitution, biological recognition, host response, and immunological detoxification. Curr Top Microbiol Immunol 216:39–81
Riviere GR, Wagoner MA, Baker-Zander SA, Weisz KS, Adams DF, Simonson L, Lukehart SA (1991) Identification of spirochetes related to Treponema pallidum in necrotizing ulcerative gingivitis and chronic periodontitis. N Engl J Med 325:539–43
Rock FL, Hardiman G, Timans JC, Kastelein RA, Bazan JF (1998) A family of human receptors structurally related to Drosophila Toll. Proc Natl Acad Sci USA 95:588–93
Rothe M, Wong SC, Henzel WJ, Goeddel DV (1994) A novel family of putative signal transducers associated with the cytoplasmic domain of the 75 kDa tumor necrosis factor receptor. Cell 78:681–92
Sato S, Nomura F, Kawai T, Takeuchi O, Muhlradt PF, Takeda K, Akira S (2000) Synergy and cross-tolerance between toll-like receptor (TLR) 2- and TLR4-mediated signaling pathways. J Immunol 165:7096–101
Schleifer KH, Kandler O (1972) Peptidoglycan types of bacterial cell walls and their taxonomic implications. Bacteriol Rev 36:407–77
Schroder NW, Opitz B, Lamping N, Michelsen KS, Zahringer U, Gobel UB, Schumann RR (2000) Involvement of lipopolysaccharide binding protein, CD14, and Toll-like receptors in the initiation of innate immune responses by Treponema glycolipids. J Immunol 165:2683–93
Schroder NW, Pfeil D, Opitz B, Michelsen KS, Amberger J, Zahringer U, Gobel UB, Schumann RR (2001) Activation of mitogen-activated protein kinases p42/44, p38, and stress-activated protein kinases in myelo-monocytic cells by Treponema lipoteichoic acid. J Biol Chem 276:9713–9
Schultz J, Milpetz F, Bork P, Ponting CP (1998) SMART, a simple modular architecture research tool: identification of signaling domains. Proc Natl Acad Sci USA 95:5857–64
Schumann RR, Leong SR, Flaggs GW, Gray PW, Wright SD, Mathison JC, Tobias PS, Ulevitch RJ (1990) Structure and function of lipopolysaccharide binding protein. Science 249:1429–31
Schwandner R, Dziarski R, Wesche H, Rothe M, Kirschning CJ (1999) Peptidoglycan- and lipoteichoic acid-induced cell activation is mediated by toll-like receptor 2. J Biol Chem 274:17406–9
Shuto T, Xu H, Wang B, Han J, Kai H, Gu XX, Murphy TF, Lim DJ, Li JD (2001) Activation of NF-κB by nontypeable Hemophilus influenzae is mediated by toll-like receptor 2-TAK1-dependent NIK-IKK alpha /beta-I κB alpha and MKK3/6-p38 MAP kinase signaling pathways in epithelial cells. Proc Natl Acad Sci USA 98:8774–9

Silverman N, Maniatis T (2001) NF-κB signaling pathways in mammalian and insect innate immunity. Genes Dev 15:2321–42

Song HY, Regnier CH, Kirschning CJ, Goeddel DV, Rothe M (1997) Tumor necrosis factor (TNF)-mediated kinase cascades: bifurcation of nuclear factor-κB and c-jun N-terminal kinase (JNK/SAPK) pathways at TNF receptor-associated factor 2. Proc Natl Acad Sci USA 94:9792–6

Takeuchi O, Hoshino K, Kawai T, Sanjo H, Takada H, Ogawa T, Takeda K, Akira S (1999) Differential roles of TLR2 and TLR4 in recognition of gram-negative and gram-positive bacterial cell wall components. Immunity 11:443–51

Takeuchi O, Hoshino K, Akira S (2000) Cutting edge: TLR2-deficient and MyD88-deficient mice are highly susceptible to Staphylococcus aureus infection. J Immunol 165:5392–6

Takeuchi O, Kawai T, Muhlradt PF, Morr M, Radolf JD, Zychlinsky A, Takeda K, Akira S (2001) Discrimination of bacterial lipoproteins by Toll-like receptor 6. Int Immunol 13:933–40

Thoma-Uszynski S, Stenger S, Takeuchi O, Ochoa MT, Engele M, Sieling PA, Barnes PF, Rollinghoff M, Bolcskei PL, Wagner M, Akira S, Norgard MV, Belisle JT, Godowski PJ, Bloom BR, Modlin RL (2001) Induction of direct antimicrobial activity through mammalian toll-like receptors. Science 291:1544–7

Ulevitch RJ, Tobias PS (1995) Receptor-dependent mechanisms of cell stimulation by bacterial endotoxin. Annu Rev Immunol 13:437–57

Underhill DM, Ozinsky A, Hajjar AM, Stevens A, Wilson CB, Bassetti M, Aderem A (1999a) The Toll-like receptor 2 is recruited to macrophage phagosomes and discriminates between pathogens. Nature 401:811–5

Underhill DM, Ozinsky A, Smith KD, Aderem A (1999b) Toll-like receptor-2 mediates mycobacteria-induced proinflammatory signaling in macrophages. Proc Natl Acad Sci USA 96:14459–63

Vabulas RM, Ahmad-Nejad P, da Costa C, Miethke T, Kirschning CJ, Hacker H, Wagner H (2001) Endocytosed HSP60 s use toll-like receptor 2 (TLR2) and TLR4 to activate the toll/interleukin-1 receptor signaling pathway in innate immune cells. J Biol Chem 276:31332–9

Visintin A, Mazzoni A, Spitzer JH, Wyllie DH, Dower SK, Segal DM (2001) Regulation of Toll-like receptors in human monocytes and dendritic cells. J Immunol 166:249–55

Vogel G (1998) Fly development genes lead to immune find. Science 281:1942–4

Wang T, Lafuse WP, Zwilling BS (2001) NFκB and Sp1 elements are necessary for maximal transcription of toll-like receptor 2 induced by Mycobacterium avium. J Immunol 167:6924–32

Werts C, Tapping RI, Mathison JC, Chuang TH, Kravchenko V, Saint Girons I, Haake DA, Godowski PJ, Hayashi F, Ozinsky A, Underhill DM, Kirschning CJ, Wagner H, Aderem A, Tobias PS, Ulevitch RJ (2001) Leptospiral lipopolysaccharide activates cells through a TLR2-dependent mechanism. Nat Immunol 2:346–52

Wesche H, Henzel WJ, Shillinglaw W, Li S, Cao Z (1997) MyD88: an adapter that recruits IRAK to the IL-1 receptor complex. Immunity 7:837–47

Williams MJ, Rodriguez A, Kimbrell DA, Eldon ED (1997) The 18-wheeler mutation reveals complex antibacterial gene regulation in Drosophila host defense. Embo J 16:6120–30

Wooten RM, Ma Y, Yoder RA, Brown JP, Weis JH, Zachary JF, Kirschning CJ, Weis JJ (2002) Toll-like receptor 2 is required for innate, but not acquired, host defense to Borrelia burgdorferi. J Immunol 168:348–55

Wright SD, Ramos RA, Tobias PS, Ulevitch RJ, Mathison JC (1990) CD14, a receptor for complexes of lipopolysaccharide (LPS) and LPS binding protein. Science 249:1431–3

Xu Y, Tao X, Shen B, Horng T, Medzhitov R, Manley JL, Tong L (2000) Structural basis for signal transduction by the Toll/interleukin-1 receptor domains. Nature 408:111–5

Xu Z, Dziarski R, Wang Q, Swartz K, Sakamoto KM, Gupta D (2001) Bacterial peptidoglycan-induced tnf-alpha transcription is mediated through the transcription factors Egr-1, Elk-1, and NF-κB. J Immunol 167:6975–82

Yang RB, Mark MR, Gray A, Huang A, Xie MH, Zhang M, Goddard A, Wood WI, Gurney AL, Godowski PJ (1998) Toll-like receptor-2 mediates lipopolysaccharide-induced cellular signalling. Nature 395:284–8

Yoshimura A, Lien E, Ingalls RR, Tuomanen E, Dziarski R, Golenbock D (1999) Cutting edge: recognition of gram-positive bacterial cell wall components by the innate immune system occurs via toll-like receptor 2. J Immunol 163:1–5

Bacterial CpG-DNA Licenses TLR9

S. BAUER and H. WAGNER

The family of Toll-like receptors (TLRs) plays an important role in the innate immune response to pathogens. TLRs sense pathogen associated molecular patterns (PAMP) and lead to the stimulation of immune cells. In man, so far ten members (TLR1–10) have been reported. This review focuses on TLR9 which is an essential component for the recognition of bacterial CpG-DNA. Expression of TLR9 and structural consideration as well as direct ligand interaction of TLR9 and CpG-DNA are discussed.

1 Introduction

In vertebrates the immune system can be broadly categorized as both an adaptive and an innate system. Adaptive immunity relies on clonally distributed T and B cells which confer a specific and memory response against pathogens. The innate immune system has developed germ-line encoded pattern recognition receptors (PRR) that promote rapid responses to microbial pathogens during the invading phase. Taking their cue from conserved pathogen-associated molecular patterns (PAMP) which are not present in the host, cells of the innate immune system activate and direct the response to pathogens (MEDZHITOV and JANEWAY 1997a).

Institute for Medical Microbiology, Hygiene and Immunology, Trogerstrasse 9, 81675 Munich, Germany, *e-mail: stefan.bauer@lrz.tum.de*

Two classes of PRR on antigen presenting cells (APC) have been identified. One class of receptor, for example the membrane anchored mannose-binding-receptors, facilitates uptake of pathogens; these receptors lead to internalization of pathogens with subsequent presentation of antigen via major histocompatibility complex molecules but do not activate APC. In contrast, toll-like receptors – members of the second class of PRR – activate APC after engagement of their cognate PAMP to express costimulatory molecules and to secrete proinflammatory and regulatory cytokines such as tumour-necrosis factor (TNF)-α, interleukin (IL)-6, IL-12 and IL-18 (Medzhitov and Janeway 2000). Overall, the innate immune system is central in sensing an infection and directing the adaptive immune response (Janeway and Medzhitov 1998).

2 Toll-Like Receptors and Their Ligands

Toll receptors are transmembrane proteins which are evolutionarily conserved between insects and vertebrates (Rock et al. 1998). In *Drosophila*, toll was first identified as essential molecule for dorsal–ventral patterning of the embryo and subsequently as key molecule for the anti-fungal immune response in the adult animal (Lemaitre et al. 1996). A homologous family of toll receptors, termed toll-like receptors (TLRs) exists in vertebrates (Rock et al. 1998). In man ten members (TLR1–10) have been reported which are fundamental in recognition of PAMP (Rock et al. 1998; Du et al. 2000; Chuang and Ulevitch 2001). TLR2 and TLR4 recognize different cell wall components of Gram-positive and Gram-negative bacteria as well as heat shock proteins (Medzhitov 2001). TLR3 and TLR5 are fundamental for the recognition of double-stranded RNA (Alexopoulou et al. 2001) and flagellin (Hayashi et al. 2001), respectively. TLR9 recognizes bacterial CpG-DNA (Hemmi et al. 2000; Bauer et al. 2001). TLR1 and TLR6 form heterodimers with TLR2 and broaden the spectrum of ligand recognition (Ozinsky et al. 2000; Hajjar et al. 2001; Takeuchi et al. 2001). The natural ligands for TLR7, TLR8 and TLR10 have not yet been identified.

3 Bacterial CpG-DNA

Studies by Tokunaga et al. first demonstrated that bacterial DNA itself was the component of bacille Calmette–Guerin which promoted immunostimulatory and anti-tumour effects (Tokunaga et al. 1984). The stimulatory effect of bacterial DNA is due to the presence of unmethylated CpG dinucleotides in a particular base context named CpG motif (Krieg et al. 1995). Vertebrate DNA is not stimulatory due to methylation of CpG dinucleotides, their low frequency (CpG suppression) and the presence of possibly inhibitory sequences (KRIEG et al. 1998). The immunostimulatory effects of bacterial DNA can be mimicked by synthetic oligodeoxynucleotides containing a CpG-motif (CpG-ODN). CpG-DNA activates macrophages and dendritic cells (DCs) to increase the expression of major histocompatibility complex and costimulatory molecules such as CD40 and CD86 on the cell surface (SPARWASSER et al. 1998). These cells also maturate and produce proinflammatory cytokines such as TNF-α, IL-1 and IL-6 as well as the regulatory cytokines IL-12 and IL-18 that promote Th1 differentiation (WAGNER 1999). The strong induction of Th1 promoting cytokines leads to a strong attenuation of Th2 biased immune responses after challenge with ragweed allergen (TIGHE et al. 2000) or schistosoma eggs (CHIARAMONTE et al. 2000). *Leishmania major* infections as a model for Th2 driven diseases can be cured by CpG-DNA treatment, even when administered as late as 20 days after infection (ZIMMERMANN et al. 1998). This strong Th1 inducing capacity of CpG-DNA also underlines its adjuvant effect during immunization leading to strong antibody responses and cytotoxic T lymphocyte induction even independently of T-cell help (LIPFORD et al. 1997a; KRIEG and WAGNER 2000; CHO et al. 2000).

4 TLR9 as a Receptor for CpG-DNA

During the search for the CpG-DNA receptor it was noted that CpG-DNA signalling is abolished in knockout mice deficient in MyD88 – an essential component in TLR signalling (see below; HACKER et al. 2000). This suggested a TLR as the receptor for CpG-DNA and in a seminal study by HEMMI et al. (2000) it was shown that macrophages from TLR9 knockout mice did not respond to CpG-DNA, but to ligands for TLR2 and TLR4. This observation was consistent with the view that TLR9 is an essential component of the CpG-DNA receptor. Subsequently it was reported that TLR9 expression in immune cells correlated with responsiveness to CpG-DNA (BAUER et al. 2001). In genetic complementation assays it was further shown that TLR9 deficient and CpG-DNA unresponsive HEK 293 cells gain responsiveness after transfection of either human or murine TLR9 cDNA (BAUER et al. 2001; TAKESHITA et al. 2001). The responsiveness was dependent on species-specific CpG-DNA motifs and explained the observation that mouse spleen cells respond better to CpG-ODN containing the core sequence GACGTT whereas human peripheral blood mononuclear cells prefer CpG-motifs containing more than one CG and the core sequence GTCGTT (KRIEG et al. 1995; BAUER et al. 1999, 2001). The response to CpG-DNA was independent of the cofactors MD-2 and CD14 which are essential components for efficient recognition of lipopolysaccharide (LPS) by TLR4 (SHIMAZU et al. 1999; BAUER et al. 2001). The hydrophilic nature of CpG-DNA in contrast with the micelle forming nature of LPS may explain the independence of TLR9 from MD-2 and CD14 although the existence of other cofactors cannot be excluded. Overall these data strongly imply that TLR9 engages CpG-DNA directly.

This hypothesis is further strengthened by the identification of two DNA-binding motifs within human and murine TLR9 which were first described in methyl-CpG-DNA binding proteins 1–4 (MBD 1–4) and in DNA methyltransferase (Hendrich and Bird 1998; Fujita et al. 2000). The DNA binding motif termed CXXC is found in proximity to the N terminus of TLR9 and mutation analysis suggests that this domain modifies the strength of protein–DNA interaction (S. Bauer et al., unpublished results). The MBD domain, directly involved in methylated and non-methylated CpG recognition within the MBD proteins 1–4, is located 228 amino acids N terminal of the transmembrane domain. The replacement of two conserved asparagine and tyrosine residues by alanines, which ablates CpG-DNA binding by MBD proteins (Fujita et al. 2000), abolishes nuclear factor (NF)-κB activation of TLR9 after CpG-DNA stimulation (S. Bauer et al., unpublished results).

Genetic complementation experiments implied direct interaction of TLR4 and LPS (Poltorak et al. 2000; Lien et al. 2000). So far, biochemical evidence for direct interaction of LPS with TLR4 (as well as MD-2 and CD14) has been provided by photo-cross linking (da Silva et al. 2001). Recently the direct interaction of peptidoglycan and TLR2 was shown by radioimmunoabsorbent assay (Murakami et al. 2001). Taken together, these results strongly imply direct interaction of PAMP with toll-like receptors. This is in contrast with *Drosophila* in which fungal infection activates a proteolytic cascade of serine proteases producing an endogenous ligand (Spätzle) that engages toll (Lemaitre et al. 1996). In vertebrates no Spätzle homologue has been identified to support the idea of a proteolytic cascade necessary for TLR activation, although a recent report claims that a serine protease is required for TLR4 function in LPS-mediated signalling (Mansell et al. 2001). It may be envisioned that LPS binding to TLR4 activates a protease cascade leading to a protein fragment analogous to Spätzle and toll in *Drosophila*. Further studies will be necessary to clarify this issue.

5 Expression of TLR9

TLR mRNA is differentially expressed and regulated in human leukocytes (Muzio et al. 2000). In particular, TLR3 expression is restricted to DCs. TLR1 is broadly expressed and can be detected in monocytes, granulocytes, T and B cells and natural killer (NK) cells. TLR2, TLR4 and TLR5 are expressed in cells of myeloid origin. TLR9 mRNA is detectable only in human B cells and CD123$^+$ plasmacytoid DCs (pDC2), whereas monocytes, and monocyte-derived DCs (MDDCs) are devoid of TLR9 mRNA (Bauer et al. 2001; Krug et al. 2001b; Kadowaki et al. 2001). This differential expression of TLR9 in DCs subtypes strongly correlates with responsiveness to CpG-DNA (Bauer et al. 2001). Plasmacytoid DCs secrete high amounts of interferon (IFN)-α after stimulation with certain CpG motifs whereas MDDCs do not respond. With LPS as microbial stimulus the activation of these subtypes is reversed: MDDCs express the LPS receptor TLR4 and respond to LPS stimulation with IL-12 and TNF-α secretion, whereas plasmacytoid DCs express only low levels of TLR4 and are not activated by LPS (Bauer et al. 2001; Miettinen et al. 2001; Krug et al. 2001b). These striking differences between MDDCs and plasmacytoid DCs in TLR expression and response to microbial stimuli suggest that these cells have evolved to recognize different pathogens. By different profiles of cytokine secretion DC subtypes may regulate the appropriate immune response against various pathogens. Accordingly, IFN-α producing plasmacytoid DCs may have evolved to control virus infections by members of virus families which contain immunostimulatory CpG-motifs.

6 Other Receptors for CpG-DNA?

It is known that different types of CpG motifs (palindromic or not) and the use of phosphodiester or phosphothioate backbone induces dramatically different profiles of cytokine release and immune activation of different cell types (Ballas et al. 1996; Verthelyi et al. 2001). For example, pCD2 cells produce high amounts of IFN-α after stimulation with a CpG-ODN containing a phosphodiester linked CpG core sequence (ODN2216) and phosphothioate modified poly G 5′ and 3′ ends (Krug et al. 2001a), but a fully phosphothioate modified CpG-ODN (ODDN2006) induces only minimal amounts of IFN-α. This raises the possibility that different receptors may recognize CpG dinucleotides in different sequence contexts and backbone composition. The requirement of TLR9 for CpG-induced signalling has been shown convincingly by TLR9 knockout studies and genetic complementation experiments for phosphothioate modified CpG-ODN and *Escherichia coli* DNA (Hemmi et al. 2000; Bauer et al. 2001). Recently, Chu et al. have suggested a role for DNA dependent protein kinase (DNA-PK) in mediating immune activation by phosphothioate protected CpG-ODN with a palindromic core sequence (Chu et al. 2000). The latter finding is inconsistent with recent reports that severe combined immunodeficient (SCID) mice that have no DNA-PK catalytic function (Blunt et al. 1996; Danska et al. 1996), respond in vivo to CpG-DNA with release of serum cytokines and succumb to TNF-driven toxic shock (Lipford et al. 1997b; Sparwasser et al. 1997). In contrast, serum cytokine release and toxic shock after in vivo CpG-DNA challenge were absent in TLR9 knock-out mice (Hemmi et al. 2000).

Genetic complementation experiments using HEK293 cells transfected with either human or murine TLR9 cDNA show that a variety of CpG-ODN with phosphodiester, phosphothiaote or mixed backbones activate NF-κB. Furthermore, murine plasmacytoid DCs secrete IFN-α after in vitro stimulation with ODN 2216 or ODN 1668 in a strictly TLR9 dependent fashion (H. Hochrein et al., manuscript in preparation). This suggests that TLR9 can act as a receptor for

CpG-ODN with different sequence contexts and backbone composition. Differences in cytokine patterns induced by these CpG-ODN may be due to varying receptor–ligand affinities or signal strength.

7 Structural Considerations for TLR9

The extracellular domain of toll-like receptors contains leucine-rich repeat (LRR) motifs which are defined by a consensus sequence of 24–29 amino acids (aa) in length and a highly conserved core region of 11 aa (LxxLxLxxN/CxL) (Kobe and Deisenhofer 1994; Kajava 1998; Rock et al. 1998). These tandem arrays of LRRs have been found in the primary structure of various proteins that participate in biological processes such as cell adhesion, signal transduction, enzyme inhibition, ribosome and DNA binding (Kobe and Deisenhofer 1994).

The crystal structure of ribonuclease inhibitor (RI), one member of the LRR protein family, reveals that LRRs correspond to structural units which consist of a β-strand and a α-helix. The structural units are arranged in such a way that all of the β-strands and the α-helices are parallel to a common axis, resulting in a horseshoe-shaped molecule with curved a parallel β-sheet lining the inner circumference of the horseshoe and the α-helices flanking its outer circumference (Kobe and Deisenhofer 1994). The number of LRRs among the family of TLR varies and for the TLR1–9 18-, 19-, 24-, 22-, 21-, 18-, 26-, 26-, 26-LRR arrays can be determined by pattern analysis. For those TLRs which contain fewer than 22 LRRs, a horseshoe-structure similar to RI can be expected. The structure of the other TLRs might be different in respect to RI, because in proteins that contain 22 or more LRRs and retain the curvature of the structure of RI, the overall organization of the repeats must be different to prevent the N and C termini from colliding. TLR9 contains 26 LRRs which are very regular compared to other TLRs. The extracellular domain can be divided into two domains formed by 14 and 12 LRRs connected through a hinge-like region (S. Bauer et al., unpublished results). It is intriguing to speculate that the CXXC domain located in domain one and the MBD motif located in domain two may associate to form a binding pocket for CpG-DNA. Crystallization and three-dimensional structure analysis of the extracellular domain of TLR9 will be important to the understanding of the structure–function relationship between CpG-DNA and TLR9.

8 TLR9-Mediated Signalling

The signalling pathways initiated by IL-1 receptor (O'Neill and Greene 1998) and TLR (Medzhitov and Janeway 1997b) are remarkably similar. The adaptor protein MyD88 is recruited to the cytoplasmic toll IL-1 receptor-like (TIR) domain presumably after ligand induced dimerization of the receptors. Recruitment of MyD88 is followed by engagement of IL-1 receptor associated kinase (IRAK) and the adaptor protein TNF receptor associated factor 6 (TRAF6). Oligomerization of TRAF6 leads to activation of downstream kinases such as the stress kinase c-Jun N-terminal kinase (JNK), p38 and IkB kinase complex resulting in activation of the transcription factors AP-1 and NF-κB.

For CpG-DNA initiated cell activation cellular uptake of DNA and endosomal maturation is required. Competition by non-stimulatory DNA or treatment of macrophages and DCs with compounds blocking endosomal maturation (chloroquine, bafilomycin) completely inhibited stimulation by CpG-DNA (Hacker et al. 1998). Signalling involves MyD88 and TRAF6 and presumably starts from endosomal like structures (Hacker et al. 2000; Schnare et al. 2000). This view is supported by analysis of RAW macrophages transfected with Myd88 tagged with green-fluorescent protein (GFP). Confocal microscopy demonstrated that GFP-MyD88 is recruited to endosome like structures upon internalization of CpG-DNA and colocalizes with TLR9 (Ahmad-Nejad et al. 2002). In contrast LPS recruits GFP-tagged MyD88 to the cell membrane where TLR4 is expressed and signalling is initiated. It was further shown that in MyD88 knockout macrophages LPS triggers TLR4 independently of MyD88 resulting in the partial activation of NF-κB, JNK and p38 (Kawai et al. 1999). This activation is mediated by the recently discovered protein termed TIR domain-containing adapter protein (TIRAP) (Fitzgerald et al. 2001; Horng et al. 2001). In contrast, TLR9 signalling relies solely on MyD88 and is independent of TIRAP (Kaisho et al. 2001; Horng et al. 2001).

9 Conclusion

The innate immune system distinguishes microbial from self DNA based on differences in CpG content and methylation status. TLR9 appears to recognize CpG-DNA directly and stimulates innate immune cells. Synthetic oligonucleotides with the CpG-motif mimic bacterial DNA and have been used successfully as therapeutic in phase II human vaccination trials and in tumour treatment studies. For the future it will be interesting to understand how CpG ODN with different sequence contexts and backbone composition control the cytokines produced by innate immune cells. This would allow optimization of CpG-DNA for therapeutic application in humans by tailoring induced cytokine patterns.

Acknowledgements. We thank G. B. Lipford for helpful discussion and H. Hochrein for critical reading of the manuscript. This work was supported by DFG-grant BA1618/2-1 and Coley Pharmaceutical Group GmbH.

References

Ahmad-Nejad P, Häcker H, Rutz M, Bauer S, Vabulas RM, Wagner H (2002) Bacterial CpG-DNA and lipopolysaccharide activate Toll-like receptors at distinct cellular compartments. Eur J Immunol, in press

Alexopoulou L, Holt AC, Medzhitov R, Flavell RA (2001) Recognition of double-stranded RNA and activation of NF-kappaB by Toll-like receptor 3. Nature 413:732–738

Ballas ZK, Rasmussen WL, Krieg AM (1996) Induction of NK activity in murine and human cells by CpG motifs in oligodeoxynucleotides and bacterial DNA. J Immunol 157:1840–1845

Bauer M, Heeg K, Wagner H, Lipford GB (1999) DNA activates human immune cells through a CpG sequence-dependent manner. Immunology 97:699–705

Bauer S, Kirschning CJ, Hacker H, Redecke V, Hausmann S, Akira S, Wagner H, Lipford GB (2001) Human TLR9 confers responsiveness to bacterial DNA via species-specific CpG motif recognition. Proc Natl Acad Sci USA 98:9237–9242

Blunt T, Gell D, Fox M, Taccioli GE, Lehmann AR, Jackson SP, Jeggo PA (1996) Identification of a nonsense mutation in the carboxyl-terminal region of DNA-dependent protein kinase catalytic subunit in the scid mouse. Proc Natl Acad Sci USA 93:10285–10290

Chiaramonte MG, Hesse M, Cheever AW, Wynn TA (2000) CpG oligonucleotides can prophylactically immunize against Th2-mediated schistosome egg-induced pathology by an IL-12-independent mechanism. J Immunol 164:973–985

Cho HJ, Takabayashi K, Cheng PM, Nguyen MD, Corr M, Tuck S, Raz E (2000) Immunostimulatory DNA-based vaccines induce cytotoxic lymphocyte activity by a T-helper cell-independent mechanism. Nat Biotechnol 18:509–514

Chu W, Gong X, Li Z, Takabayashi K, Ouyang H, Chen Y, Lois A, Chen DJ, Li GC, Karin M, Raz E (2000) DNA-PKcs is required for activation of innate immunity by immunostimulatory DNA. Cell 103:909–918

Chuang T, Ulevitch RJ (2001) Identification of hTLR10: a novel human Toll-like receptor preferentially expressed in immune cells. Biochim Biophys Acta 1518:157–161

da Silva CJ, Soldau K, Christen U, Tobias PS, Ulevitch RJ (2001) Lipopolysaccharide is in close proximity to each of the proteins in its membrane receptor complex transfer from CD14 to TLR4 and MD-2. J Biol Chem 276:21129–21135

Danska JS, Holland DP, Mariathasan S, Williams KM, Guidos CJ (1996) Biochemical and genetic defects in the DNA-dependent protein kinase in murine scid lymphocytes. Mol Cell Biol 16: 5507–5517

Du X, Poltorak A, Wei Y, Beutler B (2000) Three novel mammalian toll-like receptors: gene structure, expression, and evolution. Eur Cytokine Netw 11:362–371

Fitzgerald KA, Palsson-McDermott EM, Bowie AG, Jefferies CA, Mansell AS, Brady G, Brint E, Dunne A, Gray P, Harte MT, McMurray D, Smith DE, Sims JE, Bird TA, O'Neill LA (2001) Mal (MyD88-adapter-like) is required for Toll-like receptor-4 signal transduction. Nature 413:78–83

Fujita N, Shimotake N, Ohki I, Chiba T, Saya H, Shirakawa M, Nakao M (2000) Mechanism of transcriptional regulation by methyl-CpG binding protein MBD1. Mol. Cell Biol 20:5107–5118

Hacker H, Mischak H, Miethke T, Liptay S, Schmid R, Sparwasser T, Heeg K, Lipford GB, Wagner H (1998) CpG-DNA-specific activation of antigen-presenting cells requires stress kinase activity and is preceded by non-specific endocytosis and endosomal maturation. EMBO J 17:6230–6240

Hacker H, Vabulas RM, Takeuchi O, Hoshino K, Akira S, Wagner H (2000) Immune Cell Activation by Bacterial CpG-DNA through Myeloid Differentiation Marker 88 and Tumor Necrosis Factor Receptor-Associated Factor (TRAF)6. J Exp Med 192:595–600

Hajjar AM, O'Mahony DS, Ozinsky A, Underhill DM, Aderem A, Klebanoff SJ, Wilson CB (2001) Cutting edge: functional interactions between toll-like receptor (TLR) 2 and TLR1 or TLR6 in response to phenol-soluble modulin. J Immunol 166:15–19

Hayashi F, Smith KD, Ozinsky A, Hawn TR, Yi EC, Goodlett DR, Eng JK, Akira S, Underhill DM, Aderem A (2001) The innate immune response to bacterial flagellin is mediated by Toll-like receptor 5. Nature 410:1099–1103

Hemmi H, Takeuchi O, Kawai T, Kaisho T, Sato S, Sanjo H, Matsumoto M, Hoshino K, Wagner H, Takeda K, Akira S (2000) A Toll-like receptor recognizes bacterial DNA. Nature 408:740–745

Hendrich B, Bird A (1998) Identification and characterization of a family of mammalian methyl-CpG binding proteins. Mol Cell Biol 18:6538–6547

Horng T, Barton GM, Medzhitov R (2001) TIRAP: an adapter molecule in the Toll signaling pathway. Nat Immunol 2:835–841
Janeway CA Jr, Medzhitov R (1998) Introduction: the role of innate immunity in the adaptive immune response. Semin Immunol 10:349–350
Kadowaki N, Ho S, Antonenko S, Malefyt RW, Kastelein RA, Bazan F, Liu YJ (2001) Subsets of human dendritic cell precursors express different toll-like receptors and respond to different microbial antigens. J Exp Med 194:863–869
Kaisho T, Takeuchi O, Kawai T, Hoshino K, Akira S (2001) Endotoxin-induced maturation of MyD88-deficient dendritic cells. J Immunol 166:5688–5694
Kajava AV (1998) Structural diversity of leucine-rich repeat proteins. J Mol Biol 277:519–527
Kawai T, Adachi O, Ogawa T, Takeda K, Akira S (1999) Unresponsiveness of MyD88-deficient mice to endotoxin. Immunity 11:115–122
Kobe B, Deisenhofer J (1994) The leucine-rich repeat: a versatile binding motif. Trends Biochem Sci 19:415–421
Krieg AM, Wagner H (2000) Causing a commotion in the blood: immunotherapy progresses from bacteria to bacterial DNA. Immunol Today 21:521–526
Krieg AM, Wu T, Weeratna R, Efler SM, Love-Homan L, Yang L, Yi AK, Short D, Davis HL (1998) Sequence motifs in adenoviral DNA block immune activation by stimulatory CpG motifs. Proc Natl Acad Sci USA 95:12631–12636
Krieg AM, Yi AK, Matson S, Waldschmidt TJ, Bishop GA, Teasdale R, Koretzky GA, Klinman DM (1995) CpG motifs in bacterial DNA trigger direct B-cell activation. Nature 374:546–549
Krug A, Rothenfusser S, Hornung V, Jahrsdorfer B, Blackwell S, Ballas ZK, Endres S, Krieg AM, Hartmann G (2001a) Identification of CpG oligonucleotide sequences with high induction of IFN-alpha/beta in plasmacytoid dendritic cells. Eur. J Immunol 31:2154–2163
Krug A, Towarowski A, Britsch S, Rothenfusser S, Hornung V, Bals R, Giese T, Engelmann H, Endres S, Krieg AM, Hartmann G (2001b) Toll-like receptor expression reveals CpG DNA as a unique microbial stimulus for plasmacytoid dendritic cells which synergizes with CD40 ligand to induce high amounts of IL-12. Eur J Immunol 31:3026–3037
Lemaitre B, Nicolas E, Michaut L, Reichhart JM, Hoffmann JA (1996) The dorsoventral regulatory gene cassette spätzle/Toll/cactus controls the potent antifungal response in Drosophila adults. Cell 20; 86:973–983
Lien E, Means TK, Heine H, Yoshimura A, Kusumoto S, Fukase K, Fenton MJ, Oikawa M, Qureshi N, Monks B, Finberg RW, Ingalls RR, Golenbock DT (2000) Toll-like receptor 4 imparts ligand-specific recognition of bacterial lipopolysaccharide. J Clin Invest 105:497–504
Lipford GB, Bauer M, Blank C, Reiter R, Wagner H, Heeg K (1997a) CpG-containing synthetic oligonucleotides promote B and cytotoxic T cell responses to protein antigen: a new class of vaccine adjuvants. Eur J Immunol 27:2340–2344
Lipford GB, Sparwasser T, Bauer M, Zimmermann S, Koch ES, Heeg K, Wagner H (1997b) Immunostimulatory DNA: sequence-dependent production of potentially harmful or useful cytokines. Eur J Immunol 27:3420–3426
Mansell A, Reinicke A, Worrall DM, O'Neill LA (2001) The serine protease inhibitor antithrombin III inhibits LPS-mediated NF-kappaB activation by TLR-4. FEBS Lett 508:313–317
Medzhitov R (2001) Toll-like receptors and innate immunity. Nature reviews 1:135–145
Medzhitov R, Janeway C Jr (2000) Innate immunity. N Engl J Med 343:338–344
Medzhitov R, Janeway CA Jr (1997a) Innate immunity: impact on the adaptive immune response. Curr Opin Immunol 9:4–9
Medzhitov R, Janeway CA Jr (1997b) Innate immunity: the virtues of a nonclonal system of recognition. Cell 91:295–298
Miettinen M, Sareneva T, Julkunen I, Matikainen S (2001) IFNs activate toll-like receptor gene expression in viral infections. Genes Immun 2:349–355
Murakami S, Iwaki D, Mitsuzawa H, Sano H, Takahashi H, Voelker DR, Akino T, Kuroki Y (2001) Surfactant protein A inhibits peptidoglycan-induced TNF-alpha secretion in U937 cells and alveolar macrophages by direct interaction with toll-like receptor 2. J Biol Chem in press
Muzio M, Bosisio D, Polentarutti N, D'amico G, Stoppacciaro A, Mancinelli R, van't Veer C, Penton-Rol G, Ruco LP, Allavena P, Mantovani A (2000) Differential expression and regulation of toll-like receptors (TLR) in human leukocytes: selective expression of TLR3 in dendritic cells. J Immunol 164:5998–6004
O'Neill LA, Greene C (1998) Signal transduction pathways activated by the IL-1 receptor family: ancient signaling machinery in mammals, insects, and plants. J Leukoc Biol 63:650–657

Ozinsky A, Underhill DM, Fontenot JD, Hajjar AM, Smith KD, Wilson CB, Schroeder L, Aderem A (2000) The repertoire for pattern recognition of pathogens by the innate immune system is defined by cooperation between Toll-like receptors. Proc Natl Acad Sci USA 25; 97:13766–13771

Poltorak A, Ricciardi-Castagnoli P, Citterio S, Beutler B (2000) Physical contact between lipopolysaccharide and toll-like receptor 4 revealed by genetic complementation. Proc Natl Acad Sci USA 97:2163–2167

Rock FL, Hardiman G, Timans JC, Kastelein RA, Bazan JF (1998) A family of human receptors structurally related to Drosophila Toll. Proc Natl Acad Sci USA 20; 95:588–593

Schnare M, Holtdagger AC, Takeda K, Akira S, Medzhitov R (2000) Recognition of CpG DNA is mediated by signaling pathways dependent on the adaptor protein MyD88. Curr Biol 10:1139–1142

Shimazu R, Akashi S, Ogata H, Nagai Y, Fukudome K, Miyake K, Kimoto M (1999) MD-2, a molecule that confers lipopolysaccharide responsiveness on Toll-like receptor 4. J Exp Med 189:1777–1782

Sparwasser T, Koch ES, Vabulas RM, Heeg K, Lipford GB, Ellwart JW, Wagner H (1998) Bacterial DNA and immunostimulatory CpG oligonucleotides trigger maturation and activation of murine dendritic cells. Eur J Immunol 28:2045–2054

Sparwasser T, Miethke T, Lipford G, Erdmann A, Hacker H, Heeg K, Wagner H (1997) Macrophages sense pathogens via DNA motifs: induction of tumor necrosis factor-alpha-mediated shock. Eur J Immunol 27:1671–1679

Takeshita F, Leifer CA, Gursel I, Ishii KJ, Takeshita S, Gursel M, Klinman DM (2001) Cutting edge: role of toll-like receptor 9 in cpg dna-induced activation of human cells. J Immunol 167:3555–3558

Takeuchi O, Kawai T, Muhlradt PF, Morr M, Radolf JD, Zychlinsky A, Takeda K, Akira S (2001) Discrimination of bacterial lipoproteins by Toll-like receptor 6. Int Immunol 13:933–940

Tighe H, Takabayashi K, Schwartz D, Van Nest G, Tuck S, Eiden JJ, Kagey-Sobotka A, Creticos PS, Lichtenstein LM, Spiegelberg HL, Raz E (2000) Conjugation of immunostimulatory DNA to the short ragweed allergen amb a 1 enhances its immunogenicity and reduces its allergenicity. J Allergy Clin Immunol 106:124–134

Tokunaga T, Yamamoto H, Shimada S, Abe H, Fukuda T, Fujisawa Y, Furutani Y, Yano O, Kataoka T, Sudo T (1984) Antitumor activity of deoxyribonucleic acid fraction from Mycobacterium bovis BCG. I. Isolation, physicochemical characterization, and antitumor activity. J Natl Cancer Inst 72:955–962

Verthelyi D, Ishii K, Gursel M, Takeshita F, Klinman D (2001) Human peripheral blood cells differentially recognize and respond to two distinct CPG motifs. J Immunol 166:2372–2377

Wagner H (1999) Bacterial CpG DNA activates immune cells to signal infectious danger. Adv Immunol 73:329–368

Zimmermann S, Egeter O, Hausmann S, Lipford GB, Rocken M, Wagner H, Heeg K (1998) CpG oligodeoxynucleotides trigger protective and curative Th1 responses in lethal murine leishmaniasis. J Immunol 160:3627–3630

MyD88 as a Bottle Neck in Toll/IL-1 Signaling

O. TAKEUCHI and S. AKIRA

Myeloid differentiation factor 88 (MyD88) is an adaptor molecule composed of an N-terminal death domain and a C-terminal Toll/interleukin (IL)-1R homology domain. Ligand binding to Toll-like receptor (TLR)/IL-1R family members results in the association of MyD88 to the cytoplasmic tail of receptors; this then initiates the signaling cascade that leads to the activation of nuclear factor-κB and mitogen-activated protein kinases. Analysis of MyD88-deficient mice revealed its essential role in TLR/IL-1R signaling as well as in both the innate and the adaptive immune response.

1 Introduction

Invasion of pathogens induces host defense responses for eliminating the microorganisms. In the initial stage of infection, innate immune cells such as macrophages and dendritic cells (DCs) play a central role in the anti-microbial immune response. They produce a set of proinflammatory cytokines such as tumor necrosis factor (TNF)-α, interleukin (IL)-1β, IL-6 and IL-12 and chemokines to give rise to the inflammatory responses. Activated macrophages are also involved in the phagocytosis and direct killing of pathogens. On the other hand, DCs process

Department of Host Defense, Research Institute for Microbial Diseases, Osaka University, 3-1 Yamada-oka, Suita, Osaka 565-0871, Japan, *e-mail: otaake@biken.osaka-u.ac.jp*

pathogens, present the antigenic peptides to naive T cells together with surface co-stimulatory molecules, and activate the T cells to proliferate and differentiate. Acquired immunity, mediated by T cells having a rearranged receptor specific for the antigen of a particular microorganism, plays an essential role in the immune response in the late stage of infection (BANCHEREAU and STEINMAN 1998). Moreover, re-infection with the same pathogen results in rapid elimination by the adaptive immune system.

Host innate immune cells recognize specific components common to a range of bacteria, so called pathogen-associated molecular patterns (PAMPs) or bacterial modulins, by germline-encoded pattern recognition receptors (ADEREM and ULEVITCH 2000; AKIRA et al. 2001; MEDZHITOV and JANEWAY 1997). Macrophages and DCs exposed to PAMPs produce proinflammatory cytokines and acquire the ability to drive the development of T helper 1 (Th1) cells. Recent extensive studies revealed that the Toll-like receptor (TLR) family is critical for the recognition of these PAMPs. TLR2 recognizes peptidoglycan (PGN), lipoarabinomannan and lipoproteins (ALIPRANTIS et al. 1999; BRIGHTBILL et al. 1999; MEANS et al. 1999; TAKEUCHI et al. 1999, 2000b; YOSHIMURA et al. 1999). In contrast, TLR4 is essential for the recognition of lipopolysaccharide (LPS) of Gram-negative bacteria (HOSHINO et al. 1999; POLTORAK et al. 1998). TLR9 is critical for the response to bacterial unmethylated DNA with a CpG motif (CpG-DNA) (HEMMI et al. 2000). TLR3 was reported to be involved in the recognition of viral double-stranded RNA (ALEXOPOULOU et al. 2001). TLR6 has the unique function of discriminating subtle differences between some ligands of TLR2 and recognizes mycoplasmal di-acyl lipopeptide cooperatively with TLR2 (OZINSKY et al. 2000; TAKEUCHI et al. 2001).

Stimulation with all known TLR ligands activates common intracellular signaling cascades leading to nuclear translocation of a rel-type transcription factor nuclear factor (NF)-κB and activation of mitogen-activated protein (MAP) kinases to induce the gene expression of proinflammatory cytokines. TLRs have a modular organization consisting of extracellular leucine-rich repeat (LRR) motifs separated by a transmembrane domain from a cytoplamic region similar to that found in the IL-1R family, named the Toll/IL-1R (TIR) homology domain (Fig. 1).

The IL-1R family is defined by an extracellular immunoglobulin-like domain and the cytoplasmic TIR domain. This family includes IL-1R, IL-18R, T1/ST2 and some other orphan receptors (MITCHAM et al. 1996; PARNET et al. 1996). IL-1 is a critical mediator of the inflammatory response as it induces several genes. On the other hand, IL-18, a cytokine structurally related to IL-1, has the ability to induce Th1-polarizing cells to produce interferon (IFN)-γ and to activate natural killer cells (NAKANISHI et al. 2001). A role of IL-18 in the development of naive T cells into Th2 cells has also been reported. Stimulation with IL-1 family cytokines also results in the activation of NF-κB and c-Jun N-terminal kinase (JNK), suggesting that the TIR domain is important for triggering the common signaling cascade.

An adaptor protein – myeloid differentiation factor 88 (MyD88) – is also known to possess a TIR domain. Recent studies revealed that MyD88 is essential

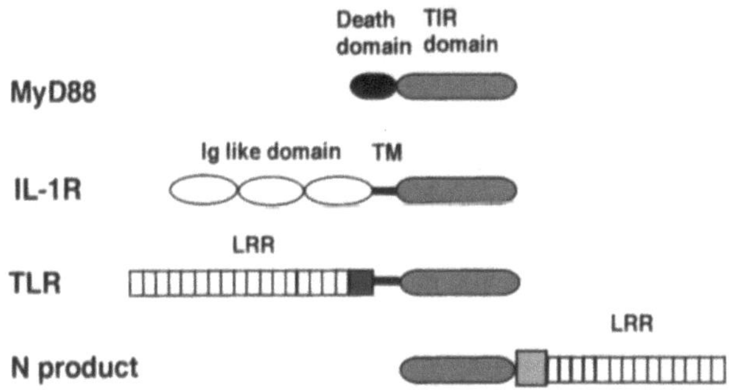

Fig. 1. The TIR domain-containing molecules. This domain is shared by various molecules that participate in the innate immune response. MyD88 consists of the N-terminal DD and the C-terminal TIR domain. A TLR is characterized by N-terminal LRRs followed by the transmembrane domain and the cytoplasmic TIR domain. Immunoglobulin-like modules in the IL-1R family distinguish it from TLR. Plant tobacco disease resistance gene products also contain this domain

for the signaling of IL-1R/TLR family members. This review will attempt to summarize recent progress in the study of the functional roles of MyD88 and the TLR signaling pathway.

2 Structure of MyD88

MyD88 was originally identified as a myeloid differentiation primary response gene activated in mouse M1 myeloid precursors following IL-6-induced terminal differentiation (Lord et al. 1990). Since then, human and mouse MyD88 mRNA have been detected in several cell lines and tissues, but not in the brain (Bonnert et al. 1997; Burns et al. 1998). The human MyD88 gene is encoded by five exons and maps to chromosome 3p21.3-p22. The mouse and human MyD88 cDNAs predict proteins of 296 amino acids. Human MyD88 shares 81% amino acid identity with the mouse counterpart. The N-terminal approximately 90 amino acids of MyD88 contain a death domain (DD) followed by the C-terminal TIR domain. The DD was initially defined as the region of similarity between the cytoplasmic domains of the Fas and TNF receptors required for their induction of cytotoxic signaling (Itoh and Nagata 1993; Weber and Vincenz 2001). Now it is known that huge numbers of proteins contain this motif and DD mediates protein–protein interactions with other DD sequences. IL-1R associated kinase (IRAK) also possesses a DD in its N terminus (Cao et al. 1996).

The C-terminal TIR domain consists of approximately 130 amino acids. As described above, it is shared by both IL-1R and TLR family members. In addition to mammals, plant disease-resistance proteins from tobacco, thale cress and flax also have these protein modules (Baker et al. 1997). They consist of an N-terminal TIR domain, a central nucleotide binding domain and C-terminal LRRs.

3 Role of MyD88 in IL-1/TLR Signaling

In 1997, it was reported that MyD88 was involved in IL-1 signaling as an adaptor protein (BURNS et al. 1998; MUZIO et al. 1997; WESCHE et al. 1997). Cao and co-workers purified a protein co-precipitated with IL-1R upon ligand stimulation in 293 cells and identified MyD88 in the complex (WESCHE et al. 1997). MyD88 binds to both IRAK and the IL-1R complex and mediates the association of IRAK with the receptor. A kinase similar to IRAK, IRAK-2, was also reported to interact with MyD88 in response to stimulation with IL-1 (MUZIO et al. 1997). Over-expression of MyD88 induced the activation of NF-κB and JNK. In contrast, the C-terminal region of MyD88 acted as a dominant-negative inhibitor of IL-1R to induce NF-κB activity. IL-18, a cytokine structurally related to IL-1, could activate NF-κB and JNK via IL-18R. IL-18R was originally cloned as an orphan IL-1R family member, named IL-1R related protein. Expression of the dominant-negative version of MyD88 also resulted in the inhibition of IL-18-induced NF-κB and JNK activation (ADACHI et al. 1998). It is possible that MyD88 participates in the signaling of other orphan receptors in the IL-1R family such as T1/ST2 and IL-1Rrp2.

In addition to the IL-1R family, TLR signaling also requires MyD88 (Fig. 2). The dominant-negative version of MyD88 inhibited TLR4-induced activation of NF-κB (MEDZHITOV et al. 1998; MUZIO et al. 1998). Phosphorylation of JNK as well as the activation of AP-1 was also inhibited. Aderem and co-workers showed that expression of dominant-negative MyD88 in mouse macrophage RAW cells resulted in the abolishment of TNF-α production in response to Gram-positive bacteria, Gram-negative bacteria, yeast and mycobacteria (UNDERHILL et al. 1999a,b).

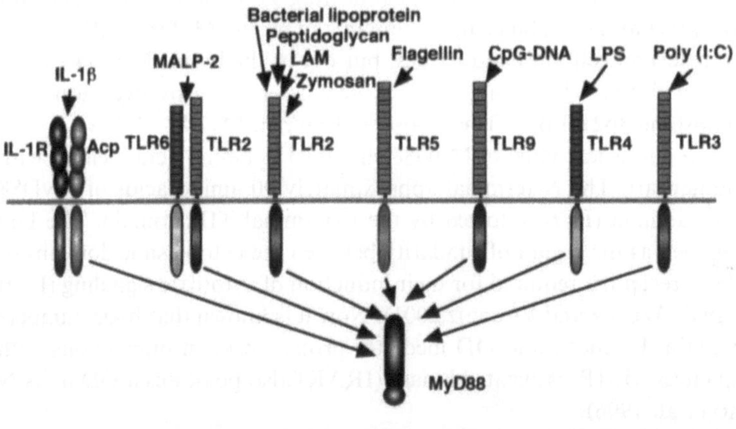

Fig. 2. MyD88 is essential for the signaling of various TLR and IL-1R. TLR2 and TLR6 cooperatively recognize MALP-2. Other TLR2 ligands such as PGN, zymosan, lipoarabinomannan and lipoproteins from various bacteria may have another unknown partner(s). In contrast, TLR4 is required for the signaling of LPS from Gram-negative bacteria. TLR3, TLR5 and TLR9 are responsible for the recognition of poly (I:C), flagellin and CpG-DNA, respectively. These PAMPs induces proinflammatory cytokines via MyD88

Further, it was reported that some viral proteins disturb the MyD88 pathway to evade the host immune system (Bowie et al. 2000). Vaccinia virus proteins, termed A46R and A52R, have been shown to share amino acid sequence similarity with the TIR domain of MyD88. It was also shown that over-expression of A46R and A52R inhibited IL-1 or TLR4-mediated NF-κB activation, by mimicking the effect of a dominant-negative version of MyD88.

4 Signaling Pathway Downstream of MyD88

Extensive studies were performed to clarify the signaling pathways coupled with the nuclear translocation of NF-κB and the activation of MAP kinases such as JNK and p38 downstream of MyD88 (Fig. 3). Upon ligand stimulation, recruitment of MyD88 to the IL-1R complex or TLR occurs, which triggers autophosphorylation of IRAK (Cao et al. 1996). Before stimulation with ligand, IRAK is captured by a recently identified molecule Tollip, and the IRAK–Tollip complex is recruited to the activated IL-1R complex, where the dissociation of IRAK from Tollip occurs (Burns et al. 2000). Over-expression of Tollip suppresses IL-1β-, TLR2- and TLR4-mediated NF-κB activation (Bulut et al. 2001). Activated IRAK released from the receptor binds and activates TNF receptor-associated factor 6 (TRAF6) to stimulate the IκB kinase (IKK) complex and MAP kinase. Experiments with TRAF6-deficient mice disclosed that TRAF6 is essential for cellular responses to IL-1, LPS, CD40 ligand and RANK (receptor activator of NF-κB) (Lomaga et al.

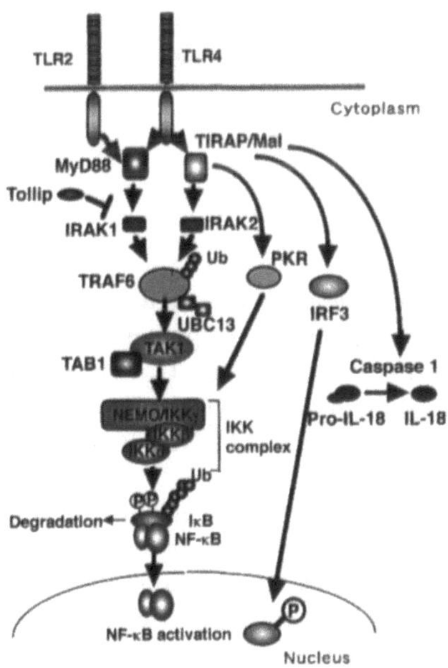

Fig. 3. TLR signaling pathways. The presence of bacterial components triggers the sequential activation of a signaling cascade leading to the nuclear translocation of NF-κB. MALP-2-induced NF-κB activation is fully MyD88-dependent, whereas LPS activates both MyD88-dependent and -independent pathways. The MyD88-independent pathway activates TIRAP/MAL. The molecule was reported to activate PKR and/or IRAK2. In addition, LPS induces nuclear translocation of IRF-3 and activation of caspase 1 to process IL-18 in a MyD88-independent manner

1999; NAITO et al. 1999). Recently, Chen and co-workers identified two activator complexes essential for TRAF6-induced IKK activation, named TRIKA1 and 2 (WANG et al. 2001). TRIKA1 consists of two ubiquitination proteins, Ubc13 and Uev1A (DENG et al. 2000). Interestingly, although polyubiquitination usually leads to proteasome-mediated protein degradation, the ubiquitination via this Ubc complex does not. The other activator, TRIKA2, consists of a complex of transforming growth factor (TGF)β-activating kinase 1 (TAK1) and associated proteins, TAB1 and TAB2 (WANG et al. 2001). It was also shown that ubiqutinated TRAF6 activates the TAK1 complex to phosphorylate the IKK complex directly. The activated IKK complex phosphorylates IκB, leading to proteasome-mediated degradation of IκB, and then NF-κB translocates from the cytosol to the nucleus to mediate the expression of proinflammatory cytokines.

Furthermore, it was reported that atypical protein kinase C (PKC) and its interacting protein, p62, were also involved in the IL-1 signaling pathway by interacting with TRAF6 to activate the IKK complex (SANZ et al. 2000). Recent studies in knockout animals disclosed a requirement of the PKC family for the activation of NF-κB-dependent gene expression. Leitges et al. generated ζPKC-deficient mice and showed that it is critical for κB-dependent transcriptional activity as well as the phosphorylation of NF-κB p65 (LEITGES et al. 2001). They also showed that TNF-α- and IL-1-induced IKK activation was attenuated in ζPKC-deficient lung, whereas it was not impaired in embryonic fibroblasts. On the other hand, Castrillo et al. established PKCε-deficient mice and demonstrated that macrophages from these mice exhibited severely reduced proinflammatory cytokine production in response to LPS (CASTRILLO et al. 2001). Activation of the IKK complex as well as nuclear translocation of NF-κB was also impaired in PKCε-deficient macrophages.

These reports imply that the signaling molecules responsible for the activation of the IKK complex differ according to cell type. Further exploration will clarify the link between TRAF6 and the IKK complex in IL-1/TLR signaling.

5 MyD88-Deficient Mice

The *MyD88* gene was disrupted in mice by homologous recombination (ADACHI et al. 1998). MyD88-deficient mice lacked T-cell proliferative responses and acute phase protein production in response to IL-1. In addition, IL-18-mediated production of IFN-γ by Th1 cells and the activation of natural killer cells were also abrogated in MyD88-deficient mice. Accordingly, activation of NF-κB and JNK was diminished in MyD88-deficient T cells.

MyD88 is critical for the signaling of not only the IL-1R family, but also of various TLRs. We observed that macrophages derived from MyD88-deficient mice did not secrete TNF-α, IL-6 or nitric oxide in response to LPS, a TLR4 ligand (KAWAI et al. 1999). The mice were highly resistant to LPS-induced shock. TLR2 ligands such as PGN and macrophage-activating lipopeptide 2kDa (MALP-2) and the TLR9 ligand CpG-DNA also failed to activate MyD88-deficient macrophages (Hacker et al. 2000; Takeuchi et al. 2000b,c). Therefore, we concluded that MyD88 is an essential mediator for the signaling of TLR2, TLR4, TLR9 and the IL-1R family. Treatment of MyD88-deficient mice with a TLR5 ligand, flagellin, failed to produce IL-6 (Hayashi et al. 2001). Recently, Flavell and co-workers showed that TLR3 participates in the recognition of viral double-stranded RNA and poly riboinosinic polyribocytidylic acid [Poly (I:C)] (Alexopoulou et al. 2001). They also showed that MyD88-deficient DCs failed to produce IL-12 and IL-6. In our laboratory, poly (I:C)-induced IFN-β mRNA expression was normal in MyD88-deficient fibroblasts compared to wild-type cells (unpublished results). Further investigation and a cautious interpretation of the results will be required to elucidate the role of the TLR-MyD88 pathway in the recognition of poly (I:C). These observations indicate that MyD88 is essential for the cytokine production induced by all known TLR ligands. In addition to the known TLR ligands, TNF-α production in response to heat-killed group B streptococcus was also abrogated in MyD88-deficient macrophages, whereas TLR2-, TLR4-, TLR6- and TLR9-deficient macrophages produced TNF-α normally, suggesting that the bacteria are recognized by TLR(s) other than those for which ligands are known (Henneke et al. 2001). Furthermore, Adachi et al. showed that the liver injury and IL-12 production caused by infection with *Plasmodium berghei* was impaired in MyD88-deficient mice, suggesting that the TLR-MyD88 signaling pathway is critical for the recognition of Malaria parasites (Adachi et al. 2001). However, TLR2, TLR4 and TLR6-deficient mice showed normal responses to *P. berghei* infection.

Furthermore, it was also shown that MyD88 plays a central role in anti-bacterial host defense. MyD88-deficient macrophages did not produce TNF-α and IL-6 in response to heat-killed whole *Staphylococcus aureus* (Takeuchi et al. 2000a). Accordingly, MyD88-deficient mice were highly susceptible to *S. aureus* infection. After intravenous infection with 1×10^7 *S. aureus* organisms, all of the MyD88-deficient mice died within 4 days, whereas all of the wild-type mice survived. Although TLR2-deficient mice were also susceptible to *S. aureus* infection, the phenotype of MyD88-deficient mice was much more severe than that of TLR2-deficient mice. The quick death of MyD88-deficient mice after inoculation with bacteria suggests that MyD88 is critical for the elimination of invading bacteria in the early stage of infection prior to the development of acquired immunity.

To investigate the role of MyD88 in antigen-specific T-cell responses, MyD88-deficient mice were immunized with ovalubumin mixed with complete Freund's adjuvant (Schnare et al. 2001). T cells obtained from the draining lymph nodes of MyD88-deficient mice did not proliferate or produce IFN-γ in response to ovalubumin stimulation, indicating that the adjuvant effect of bacterial components is mediated through a MyD88-dependent signaling pathway. Therefore, MyD88-dependent signaling is critical not only for innate immunity but also for an optimal acquired immunity.

In summary, MyD88 is essential for the production of proinflammatory cytokines following stimulation with various PAMPs, and the MyD88-dependent signaling pathway plays a central role in both innate and acquired immunity.

6 MyD88-Independent Signaling Pathway

In vitro experiments predicted that MyD88 was an adaptor protein that mediated the association between TLR and IRAK. Consistent with this notion, MALP-2- or CpG-DNA-stimulated MyD88-deficient macrophages failed activate signaling molecules such as IRAK, NF-κB and JNK (Takeuchi et al. 2000b; O. Takeuchi and S. Akira, unpublished results). However, further investigation revealed that LPS-induced activation of neither NF-κB nor JNK was abrogated in MyD88-deficient macrophages, though it was delayed (Kawai et al. 1999). In contrast, only the activation of IRAK was absent in MyD88-deficient cells. As TLR4-deficient cells failed to activate both NF-κB and JNK in response to LPS, the MyD88-independent pathway seemed to originate from TLR4. Surprisingly, this delay in the activation of transcription factors did not lead to the transcription of proinflammatory cytokine genes. These results imply that activation of additional mediators may be required for the transcription of cytokine genes. In addition, LPS induced a delayed NF-κB activation in TRAF6-deficient fibroblasts as well as MyD88/TRAF6-deficient cells, indicating that these pathways are MyD88- and TRAF6-independent (Kawai et al. 2001).

To explore the function of the MyD88-independent signaling pathway, a subtractive hybridization screening was performed (Kawai et al. 2001). The screening had identified genes induced in response to LPS in MyD88-deficient macrophages including *IP-10*, *GARG16* and *IRG1*. These genes are known to be induced in response to IFN stimulation, and are called IFN-inducible genes. We found that LPS activates IFN regulatory factor (IRF) 3 leading to the induction of the genes containing an IFN-stimulated regulatory element such as IP-10 in MyD88-deficient as well as in wild-type cells. The delayed NF-κB activation was also involved in *IP-10* gene induction. In contrast, the TLR2 ligand did not activate IRF3 or induce IFN-inducible genes.

In addition to the induction of IFN-inducible genes, the MyD88-independent pathway leads to the LPS-mediated maturation of DCs (Kaisho et al. 2001). Cytokines and bacterial components are known to induce the maturation of DCs whose hallmark is the ability to activate T cells. When MyD88-deficient bone marrow DCs were cultured with LPS, they increased the amount of surface co-stimulatory molecules such as CD40, CD80 and CD86, and caused T cells to proliferate. In contrast, TLR4-deficient DCs failed to mature in response to LPS, indicating that TLR4 is an essential component for LPS-induced DC maturation.

Another example of a MyD88-independent LPS response was reported by Seki et al. (2001). Kupffer cells are known to store a precursor of IL-18 in their cytosol and LPS stimulation induced the maturation of IL-18 via cleavage with caspase-1. Seki et al. found that MyD88-deficient cells still secreted IL-18, but not IL-1 or IL-12 upon stimulation with LPS, indicating that LPS activates caspase-1 in a MyD88-independent fashion.

7 Candidate Molecules Involved in the MyD88-Independent Pathway

Recently, a novel protein containing a TIR domain and designated TIRAP/Mal has been identified (FITZGERALD et al. 2001; HORNG et al. 2001). Medzhitov and co-workers, who named this protein TIR domain-containing adaptor protein (TIRAP), showed that its over-expression in 293 cells led to the activation of NF-κB and that TIRAP associated with TLR4 and double-stranded RNA-dependent protein kinase (PKR) (HORNG et al. 2001). Further, the expression of a dominant-negative version of TIRAP inhibited LPS-induced NF-κB activation or DC maturation, whereas it did not affect CpG-DNA signaling. In contrast, O'Neill and co-workers named the same protein MyD88-adaptor-like (MAL) (FITZGERALD et al. 2001). They showed that MAL activates NF-κB via IRAK2, but not IRAK1, by forming heterodimers with MyD88. Although both reports suggested the participation of TIRAP/MAL in MyD88-independent signaling, further investigation such as an analysis of knockout mice is needed to clarify the role this protein plays in this pathway.

The other candidate for a component of a MyD88-independent signaling pathway is the double-stranded RNA-dependent protein kinase, PKR. Several reports showed that PKR-deficient embryonic fibroblasts exhibited severely impaired NF-κB activation in response to LPS, IL-1β and poly (I:C) (GOH et al. 2000; UETANI et al. 2000). However, another report described that even poly (I:C)-induced NF-κB activation was normal in PKR-deficient embryonic fibroblasts, whereas the phosphorylation of intiation factor eIF-2α was abrogated (IORDANOV et al. 2001). Further research will be needed to resolve the discrepancy in the phenotype of these knockout animals.

8 Perspectives

Recent studies disclosed that MyD88 is a key adaptor molecule for the signaling of IL-1/TLR signaling. When one compares the mammalian inflammatory response with the *Drosophila* immune response, some differences are perceived. In the *Drosophila* immune system, infection with different types of pathogens results in the production of different kinds of anti-bacterial and/or anti-fungal peptides. Moreover, the fungus-induced Toll signaling pathway and LPS-mediated immune deficiency pathway utilize completely different intracellular signaling molecules (SILVERMAN and MANIATIS 2001). In addition, it was reported that particular pathogens induce the production of specific anti-microbial peptides in *Drosophila* (LEMAITRE et al. 1997). In contrast, all known TLR ligands signal though MyD88 to produce proinflammatory cytokine such as TNF-α in mammals. It seems that the immune system of *Drosophila* is more sophisticated than the mammalian innate immune system. Why did mammals adopt a more primitive system than *Drosophila*? One explanation is the existence of acquired immunity in mammals. The activation of a common inflammatory response favors the effective escalation of local responses recruiting immune cells. Moreover, a mono-adjuvant effect may be beneficial for the development of acquired immunity.

Another issue of importance is the difference in the signaling pathway between each TLR. As described, LPS has MyD88-independent signaling pathways leading to the activation of NF-κB as well as IRF3. The activation of NF-κB in response to both MALP-2 and CpG-DNA was completely MyD88-dependent in mouse macrophages. However, there are differences in the pattern of cytokine production following MALP-2 and CpG-DNA stimulation. For example, CpG-DNA has the unique property of stimulating human plasmacytoid DC precursors to produce IFN-α (Kadowaki et al. 2001). Future studies may emphasize the difference between TLR signaling pathways.

Recently, a *Drosophila* counterpart of mammalian MyD88, designated dMyD88, was identified by Medzhitov and co-workers (Horng and Medzhitov 2001). They showed that dMyD88 associated with both Toll and the serine/threonine kinase Pelle. When expressed in S2 cells, dMyD88 activates the induction of a *Drosomycin* reporter gene, which contrasts with the inhibition of the Toll pathway by the expression of a dominant-negative dMyD88. In addition, they showed that dMyD88 also associates with another adaptor, *Drosophila* Fas-associated death domain-containing protein (dFADD), to recruit *Drosophila* caspase (Dredd). Human MyD88 was also reported to associate with FADD to induce apoptosis in response to stimulation with TLR2 ligands (Aliprantis et al. 2000). Although the physiological role of MyD88 in apoptosis is still to be clarified, MyD88 could be a junction between inflammatory response and cell death.

MyD88 plays a central role in the innate immune response and the instruction of acquired immunity. Controlling MyD88-signaling is effective in attenuating shock response in the case of severe septic infection. Activation of MyD88-signaling will also be important for an efficient vaccination as an adjuvant. MyD88 may be a potential therapeutic target for both sepsis and vaccination.

Acknowledgements. We thank E. Horita for excellent secretarial assistance. O.T. is supported by a postdoctoral fellowship from the Japan Society for the Promotion of Science.

References

Adachi K, Tsutsui H, Kashiwamura S, Seki E, Nakano H, Takeuchi O, Takeda K, Okumura K, Van Kaer L, Okamura H, Akira S, Nakanishi K (2001) *Plasmodium berghei* infection in mice induces liver injury by an IL-12- and toll-like receptor/myeloid differentiation factor 88 dependent mechanism. J Immunol 167:5928–5934

Adachi O, Kawai T, Takeda K, Matsumoto M, Tsutsui H, Sakagami M, Nakanishi K, Akira S (1998) Targeted disruption of the MyD88 gene results in loss of IL-1- and IL-18-mediated function. Immunity 9:143–150

Aderem A, Ulevitch RJ (2000) Toll-like receptors in the induction of the innate immune response. Nature 406:782–787

Akira S, Takeda K, Kaisho T (2001) Toll-like receptors: critical proteins linking innate and acquired immunity. Nat Immunol 2:675–680

Alexopoulou L, Holt AC, Medzhitov R, Flavell RA (2001) Recognition of double-stranded RNA and activation of NF-κB by Toll- like receptor 3. Nature 413:732–738

Aliprantis AO, Yang RB, Mark MR, Suggett S, Devaux B, Radolf JD, Klimpel GR, Godowski P, Zychlinsky A (1999) Cell activation and apoptosis by bacterial lipoproteins through toll-like receptor-2. Science 285:736–739

Aliprantis AO, Yang RB, Weiss DS, Godowski P, Zychlinsky A (2000) The apoptotic signaling pathway activated by Toll-like receptor-2. Embo J 19:3325–3336

Baker B, Zambryski P, Staskawicz B, Dinesh-Kumar SP (1997) Signaling in plant-microbe interactions. Science 276:726–733

Banchereau J, Steinman RM (1998) Dendritic cells and the control of immunity. Nature 392:245–252

Bonnert TP, Garka KE, Parnet P, Sonoda G, Testa JR, Sims JE (1997) The cloning and characterization of human MyD88: a member of an IL-1 receptor related family. FEBS Lett 402:81–84

Bowie A, Kiss-Toth E, Symons JA, Smith GL, Dower SK, O'Neill LA (2000) A46R and A52R from vaccinia virus are antagonists of host IL-1 and toll-like receptor signaling. Proc Natl Acad Sci U S A 97:10162–10167

Brightbill HD, Libraty DH, Krutzik SR, Yang RB, Belisle JT, Bleharski JR, Maitland M, Norgard MV, Plevy SE, Smale ST, Brennan PJ, Bloom BR, Godowski PJ, Modlin RL (1999) Host defense mechanisms triggered by microbial lipoproteins through toll-like receptors. Science 285:732–736

Bulut Y, Faure E, Thomas L, Equils O, Arditi M (2001) Cooperation of Toll-like receptor 2 and 6 for cellular activation by soluble tuberculosis factor and Borrelia burgdorferi outer surface protein A lipoprotein: role of Toll-interacting protein and IL-1 receptor signaling molecules in Toll-like receptor 2 signaling. J Immunol 167:987–994

Burns K, Clatworthy J, Martin L, Martinon F, Plumpton C, Maschera B, Lewis A, Ray K, Tschopp J, Volpe F (2000) Tollip, a new component of the IL-1RI pathway, links IRAK to the IL-1 receptor. Nat Cell Biol 2:346–351

Burns K, Martinon F, Esslinger C, Pahl H, Schneider P, Bodmer JL, Di Marco F, French L, Tschopp J (1998) MyD88, an adapter protein involved in interleukin-1 signaling. J Biol Chem 273:12203–12209

Cao Z, Henzel WJ, Gao X (1996) IRAK: a kinase associated with the interleukin-1 receptor. Science 271:1128–1131

Castrillo A, Pennington DJ, Otto F, Parker PJ, Owen MJ, Bosca L (2001) Protein kinase cepsilon is required for macrophage activation and defense against bacterial infection. J Exp Med 194: 1231–1242

Deng L, Wang C, Spencer E, Yang L, Braun A, You J, Slaughter C, Pickart C, Chen ZJ (2000) Activation of the IkappaB kinase complex by TRAF6 requires a dimeric ubiquitin conjugating enzyme complex and a unique polyubiquitin chain. Cell 103:351–361

Fitzgerald KA, Palsson-McDermott EM, Bowie AG, Jefferies CA, Mansell AS, Brady G, Brint E, Dunne A, Gray P, Harte MT, McMurray D, Smith DE, Sims JE, Bird TA, O'Neill LA (2001) Mal (MyD88-adapter-like) is required for Toll-like receptor-4 signal transduction. Nature 413:78–83

Goh KC, deVeer MJ, Williams BR (2000) The protein kinase PKR is required for p38 MAPK activation and the innate immune response to bacterial endotoxin. Embo J 19:4292–4297

Hacker H, Vabulas RM, Takeuchi O, Hoshino K, Akira S, Wagner H (2000) Immune Cell Activation by Bacterial CpG-DNA through Myeloid Differentiation Marker 88 and Tumor Necrosis Factor Receptor-Associated Factor (TRAF)6. J Exp Med 192:595–600

Hayashi F, Smith KD, Ozinsky A, Hawn TR, Yi EC, Goodlett DR, Eng JK, Akira S, Underhill DM, Aderem A (2001) The innate immune response to bacterial flagellin is mediated by Toll-like receptor 5. Nature 410:1099–1103

Hemmi H, Takeuchi O, Kawai T, Kaisho T, Sato S, Sanjo H, Matsumoto M, Hoshino K, Wagner H, Takeda K, Akira S (2000) A Toll-like receptor recognizes bacterial DNA. Nature 408:740–745

Henneke P, Takeuchi O, van Strijp JA, Guttormsen H-K, Smith JA, Schromm AB, Espevik TA, Akira S, Nizet V, Kasper D, Golenbock DT (2001) Novel engagement of CD14 and multiple toll-like receptors by group B Streptococci. J Immunol 167:7069–7076

Horng T, Barton GM, Medzhitov R (2001) TIRAP: an adapter molecule in the Toll signaling pathway. Nat Immunol 2:835–841

Horng T, Medzhitov R (2001) Drosophila MyD88 is an adapter in the Toll signaling pathway. Proc Natl Acad Sci USA 98:12654–12658

Hoshino K, Takeuchi O, Kawai T, Sanjo H, Ogawa T, Takeda Y, Takeda K, Akira S (1999) Cutting edge: Toll-like receptor 4 (TLR4)-deficient mice are hyporesponsive to lipopolysaccharide: evidence for TLR4 as the Lps gene product. J Immunol 162:3749–3752

Iordanov MS, Wong J, Bell JC, Magun BE (2001) Activation of NF-κB by double stranded RNA (dsRNA) in the absence of protein kinase R and RNase L demonstrates the existence of two separate dsRNA-triggered antiviral programs. Mol Cell Biol 21:61–72

Itoh N, Nagata S (1993) A novel protein domain required for apoptosis. Mutational analysis of human Fas antigen. J Biol Chem 268:10932–10937

Kadowaki N, Antonenko S, Liu YJ (2001) Distinct CpG DNA and polyinosinic-polycytidylic acid double-stranded RNA, respectively, stimulate CD11c-type 2 dendritic cell precursors and $CD11c^+$ dendritic cells to produce type I IFN. J Immunol 166:2291–2295

Kaisho T, Takeuchi O, Kawai T, Hoshino K, Akira S (2001) Endotoxin-induced maturation of MyD88-deficient dendritic cells. J Immunol 166:5688–5694

Kawai T, Adachi O, Ogawa T, Takeda K, Akira S (1999) Unresponsiveness of MyD88 deficient mice to endotoxin. Immunity 11:115–122

Kawai T, Takeuchi O, Fujita T, Inoue J, Muhlradt PF, Sato S, Hoshino K, Akira S (2001) Lipopolysaccharide stimulates the MyD88-independent pathway and results in activation of IFN-regulatory factor 3 and the expression of a subset of lipopolysaccharide-inducible genes. J Immunol 167: 5887–5894

Leitges M, Sanz L, Martin P, Duran A, Braun U, Garcia JF, Camacho F, Diaz-Meco MT, Rennert PD, Moscat J (2001) Targeted Disruption of the zetaPKC gene results in the impairment of the NF-κB pathway. Mol Cell 8:771–780

Lemaitre B, Reichhart JM, Hoffmann JA (1997) *Drosophila* host defense: differential induction of antimicrobial peptide genes after infection by various classes of microorganisms. Proc Natl Acad Sci USA 94:14614–14619

Lomaga MA, Yeh WC, Sarosi I, Duncan GS, Furlonger C, Ho A, Morony S, Capparelli C, Van G, Kaufman S, van der Heiden A, Itie A, Wakeham A, Khoo W, Sasaki T, Cao Z, Penninger JM, Paige CJ, Lacey DL, Dunstan CR, Boyle WJ, Goeddel DV, Mak TW (1999) TRAF6 deficiency results in osteopetrosis and defective interleukin-1, CD40, and LPS signaling. Genes Dev 13:1015–1024

Lord KA, Hoffman-Liebermann B, Liebermann DA (1990) Nucleotide sequence and expression of a cDNA encoding MyD88, a novel myeloid differentiation primary response gene induced by IL6. Oncogene 5:1095–1097

Means TK, Lien E, Yoshimura A, Wang S, Golenbock DT, Fenton MJ (1999) The CD14 ligands lipoarabinomannan and lipopolysaccharide differ in their requirement for Toll-like receptors. J Immunol 163:6748–6755

Medzhitov R, Janeway CA, Jr (1997) Innate immunity: the virtues of a nonclonal system of recognition. Cell 91:295–298

Medzhitov R, Preston-Hurlburt P, Kopp E, Stadlen A, Chen C, Ghosh S, Janeway CA, Jr. (1998) MyD88 is an adaptor protein in the hToll/IL-1 receptor family signaling pathways. Mol Cell 2: 253–258

Mitcham JL, Parnet P, Bonnert TP, Garka KE, Gerhart MJ, Slack JL, Gayle MA, Dower SK, Sims JE (1996) T1/ST2 signaling establishes it as a member of an expanding interleukin-1 receptor family. J Biol Chem 271:5777–5783

Muzio M, Natoli G, Saccani S, Levrero M, Mantovani A (1998) The human toll signaling pathway: divergence of nuclear factor κB and JNK/SAPK activation upstream of tumor necrosis factor receptor-associated factor 6 (TRAF6). J Exp Med 187:2097–2101

Muzio M, Ni J, Feng P, Dixit VM (1997) IRAK (Pelle) family member IRAK-2 and MyD88 as proximal mediators of IL-1 signaling. Science 278:1612–1615

Naito A, Azuma S, Tanaka S, Miyazaki T, Takaki S, Takatsu K, Nakao K, Nakamura K, Katsuki M, Yamamoto T, Inoue J (1999) Severe osteopetrosis, defective interleukin-1 signalling and lymph node organogenesis in TRAF6-deficient mice. Genes Cells 4:353–362

Nakanishi K, Yoshimoto T, Tsutsui H, Okamura H (2001) Interleukin-18 is a unique cytokine that stimulates both Th1 and Th2 responses depending on its cytokine milieu. Cytokine Growth Factor Rev 12:53–72

Ozinsky A, Smith KD, Hume D, Underhill DM (2000) Co-operative induction of pro inflammatory signaling by Toll-like receptors. J Endotoxin Res 6:393–396

Parnet P, Garka KE, Bonnert TP, Dower SK, Sims JE (1996) IL-1Rrp is a novel receptor-like molecule similar to the type I interleukin-1 receptor and its homologues T1/ST2 and IL-1R AcP. J Biol Chem 271:3967–3970

Poltorak A, He X, Smirnova I, Liu MY, Huffel CV, Du X, Birdwell D, Alejos E, Silva M, Galanos C, Freudenberg M, Ricciardi-Castagnoli P, Layton B, Beutler B (1998) Defective LPS signaling in C3H/HeJ and C57BL/10ScCr mice: mutations in Tlr4 gene. Science 282:2085–2088

Sanz L, Diaz-Meco MT, Nakano H, Moscat J (2000) The atypical PKC-interacting protein p62 channels NF-kappaB activation by the IL-1-TRAF6 pathway. Embo J 19:1576–1586

Schnare M, Barton GM, Holt AC, Takeda K, Akira S, Medzhitov R (2001) Toll-like receptors control activation of adaptive immune responses. Nat Immunol 2:947–950

Seki E, Tsutsui H, Nakano H, Tsuji N, Hoshino K, Adachi O, Adachi K, Futatsugi S, Kuida K, Takeuchi O, Okamura H, Fujimoto J, Akira S, Nakanishi K (2001) Lipopolysaccharide induced IL-18 secretion from murine Kupffer cells independently of myeloid differentiation factor 88 that is critically involved in induction of production of IL-12 and IL-1beta. J Immunol 166:2651–2657

Silverman N, Maniatis T (2001) NF-κB signaling pathways in mammalian and insect innate immunity. Genes Dev 15:2321–2342

Takeuchi O, Hoshino K, Akira S (2000a) Cutting edge: TLR2-deficient and MyD88-deficient mice are highly susceptible to *Staphylococcus aureus* infection. J Immunol 165:5392–5396

Takeuchi O, Hoshino K, Kawai T, Sanjo H, Takada H, Ogawa T, Takeda K, Akira S (1999) Differential roles of TLR2 and TLR4 in recognition of gram-negative and gram-positive bacterial cell wall components. Immunity 11:443–451

Takeuchi O, Kaufmann A, Grote K, Kawai T, Hoshino K, Morr M, Muhlradt PF, Akira S (2000b) Cutting edge: preferentially the R-stereoisomer of the mycoplasmal lipopeptide macrophage-activating lipopeptide-2 activates immune cells through a toll-like receptor 2- and MyD88-dependent signaling pathway. J Immunol 164:554–557

Takeuchi O, Kawai T, Muhlradt PF, Morr M, Radolf JD, Zychlinsky A, Takeda K, Akira S (2001) Discrimination of bacterial lipoproteins by Toll-like receptor 6. Int Immunol 13:933–940

Takeuchi O, Takeda K, Hoshino K, Adachi O, Ogawa T, Akira S (2000c) Cellular responses to bacterial cell wall components are mediated through MyD88-dependent signaling cascades. Int Immunol 12:113–117

Uetani K, Der SD, Zamanian-Daryoush M, de La Motte C, Lieberman BY, Williams BR, Erzurum SC (2000) Central role of double-stranded RNA-activated protein kinase in microbial induction of nitric oxide synthase. J Immunol 165:988–996

Underhill DM, Ozinsky A, Hajjar AM, Stevens A, Wilson CB, Bassetti M, Aderem A (1999a) The Toll-like receptor 2 is recruited to macrophage phagosomes and discriminates between pathogens [see comments]. Nature 401:811–815

Underhill DM, Ozinsky A, Smith KD, Aderem A (1999b) Toll-like receptor-2 mediates mycobacteria-induced proinflammatory signaling in macrophages. Proc Natl Acad Sci USA 96:14459–14463

Wang C, Deng L, Hong M, Akkaraju GR, Inoue J, Chen ZJ (2001) TAK1 is a ubiquitin dependent kinase of MKK and IKK. Nature 412:346–351

Weber CH, Vincenz C (2001) The death domain superfamily: a tale of two interfaces? Trends Biochem Sci 26:475–481

Wesche H, Henzel WJ, Shillinglaw W, Li S, Cao Z (1997) MyD88: an adapter that recruits IRAK to the IL-1 receptor complex. Immunity 7:837–847

Yoshimura A, Lien E, Ingalls RR, Tuomanen E, Dziarski R, Golenbock D (1999) Cutting edge: recognition of Gram-positive bacterial cell wall components by the innate immune system occurs via Toll-like receptor 2. J Immunol 163:1–5

Heat Shock Proteins as Ligands of Toll-Like Receptors

R.M. VABULAS[1], H. WAGNER[1], and H. SCHILD[2]

Toll-like receptors (TLRs) have been described as sensors for pathogen-associated molecular patterns crucial for the initiation of an innate immune response. These mechanisms were developed long before the adaptive immune system evolved. The latest additions to the growing list of TLR ligands are heat shock proteins (HSPs). Interestingly, not only bacterial but also mammalian HSPs interact with TLRs demonstrating that the exclusive association of TLRs with microbial ligands is obsolete. Human HSP60 and Gp96 are the first examples of non-pathogen derived ligands of TLRs. More importantly, Gp96 provides the first example of how the innate and adaptive immune system can be stimulated simultaneously by the same molecule which is released under physiological conditions from necrotic cells. Understanding the mechanisms of innate immune system interaction with HSPs will make it possible to rationally modulate immune responses, either towards immunity or towards tolerance.

1 Introduction

Heat shock proteins (HSPs) owe their designation to the observation that the synthesis of the first family members to be identified could be induced at higher temperatures than conventionally used (LINDQUIST 1986). The group of HSPs is

[1] Institute for Medical Microbiology, Immunology and Hygiene, Technical University of Munich, Trogerstr. 4a, 81675 Munich, Germany, *e-mail: martin.vabulas@erz.tum.de*
[2] Institute for Cell Biology, Department of Immunology, University of Tübingen, Auf der Morgenstelle 15, 72076 Tübingen, Germany, *e-mail: hansjoerg.schild@uni-tuebingen.de*

immense with regard to the number of nominal members in both prokaryotic and eukaryotic organisms. For a non-expert it is astonishing how different HSPs are. Most of them have nothing in common except for the name. HSPs were originally assigned to subfamilies according to their molecular weight. This simple classification criterion turned out to be a fortunate anticipation of the homologies and hence of the evolutionary relationship of proteins within a given 'size' subgroup.

HSPs merit their notoriety to the fundamental question of how proteins reach or retain their proper conformation. It is well known that the cellular environment promotes protein misfolding and aggregation. For example, the crowded cytoplasm (200–300g protein/l) reduces the diffusion coefficient and hence the mobility of macromolecules or components thereof by factors up to tenfold (Ellis 2001). In addition, equilibrium constants for macromolecular associations increase. These circumstances and the fact that exposed hydrophobic stretches can emerge before the completion of protein translation strongly favor unproductive protein folding and thus subsequent aggregation. HSPs appear to act as a counterbalance of the aggregation-prone environmental conditions by helping proteins to reach their native higher-ordered structure. The analysis of HSP functions led to the abandonment of the originally prevailing opinion claiming that protein folding is a spontaneous process and the molecular chaperone concept arose (Ellis 1987). Chaperones were defined as 'proteins that bind to and stabilize an otherwise unstable conformer of another protein – and by controlled binding and release of the substrate protein, facilitates its correct fate in vivo: be it folding, oligomeric assembly, transport to a particular subcellular compartment, or controlled switching between active/inactive conformations' (Hendrick and Hartl 1993). Many HSPs function as molecular chaperones (Table 1).

2 HSPs as Molecular Chaperones

The most widely known HSPs as far as their function is concerned are HSP60 (also called chaperonins) and HSP70. The prototype HSP60 is an *Escherichia coli* protein GroEL. GroEL is synthesized together with GroES from the same operon and is indispensable for survival even at normal temperature during both log growth and stationary phase (Fayet et al. 1989). Homologous proteins in eukaryotic cells are encoded in nuclear DNA but reside in mitochondria (HSP60 and HSP10) or chloroplasts (Cpn60 and Cpn10). A comparison of HSP60 sequences of prokaryotes and eukaryotes indicates that the nuclear encoded eukaryotic HSP60 sequences may be the result of lateral gene transfer and are probably derived from α-proteobacteria (Karlin and Brocchieri 2000). GroEL forms a double-ring complex in which each ring is composed of seven identical units (Bukau and Horwich 1998). The rings lie together, in contact with each other via equatorial closed platforms and apical openings that face in opposite directions. This

Table 1. HSP family (based on AGASHE and HARTL 2000)

HSP family	Prokaryotes (*E. coli*)		Higher eukaryotes	
	Name	Function	Name	Function
HSP10	GroES	A cofactor for GroEL	HSP10	Resident in mitochondria: a cofactor for mitochondrial HSP60
HSP40	DnaJ	A cofactor for DnaK	HSP40	Cooperates with HSC73 in nascent chain binding
HSP60	GroEL	Protein folding. Interacts with ~20% of newly synthesized proteins	HSP60	Protein folding in mitochondria
			TRiC/CCT	Protein folding in cytoplasm. Interacts with ~15% of newly synthesized proteins
HSP70	DnaK	Stress response regulation. Binds and stabilizes hydrophobic segments in unfolded proteins	HSP72	Stress-inducible. Prevents protein aggregates after stress
			HSC73	Binds nascent and newly synthesized polypeptide chains. Uncoats clathrin-coated vesicles
HSP90	HtpG	Nonessential	HSP90	Regulates activity of some signaling proteins and transcription factors
HSP100	Clp	Catalyze the unfolding, disassembly and disaggregation of proteins. Together with peptidases form ATP-dependent proteases	Clp	

impressive cellular machine weighs about one million Daltons and utilizes ATP binding and hydrolysis to drive ordered conformational changes. The central cavity of each ring functions in two major states. In the binding-active state, it is open at the end of the cylinder for ingress of substrate proteins. The folding-active state is achieved by conformational changes that are induced by the binding of ATP and by the binding of a cap-like GroES heptamer. The cavity of the bound ring enlarges about twofold in volume, giving an opportunity for a non-native protein to achieve its correct conformation in 'sparing' sequestrated space. The hydrolysis of bound ATP relaxes the high affinity interaction between the GroEL ring and the GroES cap, promoting the release of trapped protein (for functional details, see BUKAU and HORWICH 1998). In *E. coli* approximately 20% of newly synthesized polypeptides are folded by GroEL (HOURY et al. 1999). Apparently, there is no consensus sequence in substrates that bind GroEL, although preferred proteins have two or more domains with αβ-folds, which contain α-helices and buried β-sheets with extensive hydrophobic surfaces (HOURY et al. 1999).

Eubacterial GroEL and eukaryotic organellar HSP60 belong to the group I chaperonins based on evolutionary kinship and structural similarities. The group II chaperonins are represented by TRiC (TCP-1 containing ring complex), also called CCT (chaperonin containing TCP-1) from the cytosol of eukaryotes, and by thermosome, the archeal chaperonin. Proteins from this group do not use a GroES-like cofactor and form slightly different supramolecular structures. For example, TRiC contains eight distinct, albeit related, subunits per ring (Kubota et al. 1994). It was previously believed that TRiC folds only a few cytoskeletal proteins (Gao et al. 1992; Yaffe et al. 1992), but recently it was shown that up to 15% of all newly synthesized polypeptides ranging from 30 to 60 kDa in size transiently associate with TRiC (Thulasiraman et al. 1999).

HSP70s are another well studied group of heat shock proteins. HSP70s are highly conserved and show a 60%–78% identity among eukaryotes and a 40%–60% identity between eukaryotic HSP70 and *E. coli* DnaK (Kiang and Tsokos 1998). In eukaryotes, the HSP70s are found in the cytoplasm, nucleus, mitochondria, and the endoplasmic reticulum. The HSP70s are thought to play a major role in de novo protein folding by protecting nascent and newly synthesized polypeptides from aggregation. HSP70s have two domains, an N-terminal ATPase domain and a C-terminal domain divided into a conserved substrate binding part and a less conserved immediate C terminus. HSP70 chaperone activity involves cycles of polypeptide binding and release. The DnaK cycle involves: (1) a cofactor DnaJ binding to substrate; (2) substrate transfer from DnaJ to DnaK–ATP with subsequent ATP hydrolysis; (3) GrpE binding to DnaK–ADP–substrate complex and subsequent exchange of ADP to ATP; (4) release of substrate and GrpE returning DnaK to its initial state (Bukau and Horwich 1998). The eukaryotic HSP70 differs from DnaK in being independent, for its cycle, of a GrpE-like exchange factor. All HSPs studied bind short hydrophobic segments of polypeptides, thus preventing further folding or aggregation by shielding off these 'vulnerable' segments. For example, DnaK preferentially binds motifs consisting of a hydrophobic core of four to five residues particularly enriched in leucine (Rudiger et al. 1997) as determined by scanning of the cellulose-bound peptides from sequences of 37 biologically relevant proteins. HSP70 binding does not appear to induce global conformational changes in the substrate.

The biochemistry of other HSPs is not understood so deeply as that of chaperonins and HSP70s. From the immunologist's point of view, HSP90 is highly interesting because both cytoplasmic (HSP90α/β) as well as endoplasmic reticulum (Grp94/Gp96) resident forms were successfully used for vaccination (see later). HSP90 chaperones are among the most abundant proteins in the cytosol of eukaryotic cells. That HSP90 contributes to the stress response has been demonstrated by in vitro folding assays in which HSP90 acted to prevent aggregation of unfolded reporter enzymes (Mayer and Bukau 1999). HSP90 has recently been established as an ATP-dependent chaperone (Prodromou et al. 1997), but the exact mechanism of HSP90 activity is still unclear. HSP90 differs from other HSPs in that most of its defined substrates are signal transduction proteins, such as steroid hormone receptors and signaling kinases, existing in ligand-bound or ligand-free conformations (Pratt 1997).

3 HSPs and Immunity

In addition to their involvement in the control of the integrity of cellular proteins, HSPs display additional features that attracted the attention of immunologists. The pioneering observation can be traced back more than 15 years and described the induction of tumor-specific immunity upon the injection of HSP preparations into mice (SRIVASTAVA et al. 1994). Tumor-specific protection is mediated by $CD8^+$ T cells as shown by in vivo cell depletion studies and by the ability to generate cytotoxic T lymphocyte (CTL) lines specific for a variety of antigens from mice immunized with HSP molecules (ARNOLD et al. 1995; SUTO and SRIVASTAVA 1995). The HSPs able to induce these effects include HSP70, HSC70, HSP90, Gp96, Grp110 and Grp170. Experiments performed during the past years have contributed in particular to the molecular understanding of the initial observation.

The key findings can be summarized in chronological order as follows: (1) immunization with HSPs induces specific immunity against the cells from which the HSPs have been purified; (2) the specificities of the HSP-induced immune responses are based on the association of HSP with peptides generated by the degradation of intracellular proteins; (3) the major histocompatibility complex (MHC) haplotype of the cell from which the HSPs have been purified does not influence the repertoire of associated peptides; it follows that HSPs are able to induce cross-priming; (4) HSP molecules also provide an immunogenic context to synthetic peptides complexed to HSP70 or Gp96 molecules in vitro; (5) HSP-associated peptides are represented by MHC molecules after receptor-mediated endocytosis by professional antigen presenting cells (APCs); (6) the receptor responsible for endocytosis of HSPs is the α_2-macroglobulin receptor (CD91); (7) the interaction of HSPs (HSP60, HSP70, HSP90 and Gp96) with professional APCs results in the up-regulation of co-stimulatory molecules and in the release of the pro-inflammatory cytokines tumor necrosis factor (TNF)-α and interleukin (IL)-12; (8) the activation of APCs is mediated by toll-like receptor (TLR)2/TLR4-dependent signaling via the Toll/IL-1 receptor signal pathway.

3.1 HSPs Interact with Professional Antigen-Presenting Cells

To explain the immunogenicity of HSP–peptide complexes we will take a look at the basic mechanism of antigen processing and presentation. The adaptive immune system relies on the presentation of antigenic peptides bound to MHC molecules at the cell surface. Complexes of MHC molecules and peptides are recognized by T cell receptors (TCR). These interactions take place during the initial activation of naive T cells by professional APCs that display, for example, viral antigens on their MHC molecules; once activated CTLs eliminate virus-infected cells. For the activation of naive T cells, costimulatory signals (signal 2) are required in addition to TCR-mediated stimuli (signal 1). These signals are best produced by dendritic cells

(DCs), the key member of the professional APC family. DCs have been suggested to be the most sophisticated professional APC that is able to prime T cells as well as to induce tolerance by carefully regulating its arsenal of co-stimulatory surface molecules (BANCHEREAU et al. 2000).

To explain the immunogenicity of HSP–peptide complexes it was originally hypothesized that HSPs exploit the existing pathways of antigen presentation by directly targeting professional APCs able to present the associated peptides by MHC molecules (SRIVASTAVA et al. 1994). In the meantime, this hypothesis has in part been validated. Macrophages and DCs pulsed with Gp96–peptide complexes re-present the associated peptides on their MHC class I molecules and activate CTLs in vitro (SUTO and SRIVASTAVA 1995; SINGH-JASUJA et al. 2000a; CASTELLINO et al. 2000).

3.2 Receptor-Mediated Uptake of HSPs

How do HSPs interact with professional APCs? The first description of HSP–APC interaction already demonstrated specificity via receptor-mediated endocytosis (ARNOLD-SCHILD et al. 1999). In this study, the binding of gold-labeled HSC70 and Gp96 molecules to macrophages was investigated. This observation was confirmed by confocal microscopy studies (WASSENBERG et al. 1999). Evidence for binding to specific receptors was provided by flow cytometry studies using fluorescein isothiocyanate (FITC)-labeled HSC70 (CASTELLINO et al. 2000) or Gp96 (SINGH-JASUJA et al. 2000b). Saturation and competition analyses by non-labeled Gp96 demonstrated the specificity of the HSP–receptor interaction. Furthermore, only APCs such as macrophages, DCs and B cells but not T cells bound Gp96 in a receptor-mediated fashion. These studies also showed cross-presentation of HSP-associated peptide on MHC class I molecules for activation of CTL lines. Interestingly, for both molecules – Gp96 and HSC70 – receptor-mediated uptake of the HSP is essential for re-presentation of HSP-associated peptides on MHC class I molecules. Non-specific endocytosis and macropinocytosis was ineffective. Consequently, HSP receptors do not only facilitate the efficient uptake of HSP–peptide complexes which is crucial for shuttling HSP-associated antigen into the right processing pathway. For HSC70 it has also been demonstrated that the pathway chosen is dependent on the nature of the antigen and the requirement for further processing. If C-terminal processing of the peptide for transfer onto MHC molecules is required, re-presentation is dependent on the activity of the proteasome and TAP (transporter associated with antigen processing). If C-terminal processing is not required, transfer of peptide from HSP to MHC can occur inside the endosome as confirmed by confocal microscopy (CASTELLINO et al. 2000).

The search for the identity of one or more HSP receptors is still in progress. The first receptor for Gp96 has been identified as CD91 by cross-linking Gp96 chemically to plasma membrane fractions of macrophages and sequencing the bound protein (BINDER et al. 2000a). Meanwhile CD91 has also been identified as a receptor for the heat shock proteins HSP70, HSP90 and Calreticulin (BASU et al.

2001). CD91, also known as α_2-macroglobulin receptor or low density lipoprotein receptor-related protein, consists of a 420-kDa α- and a 85-kDa β-unit. Its functions are believed to be the clearance of extracellular plasma products and the regulation of lipid metabolism (reviewed by STRICKLAND et al. 1995). The question of whether CD91 is the only receptor for Gp96 or whether CD91 is also a receptor for other HSPs still needs to be resolved.

Interestingly, the classical ligand of CD91, the plasma protein α_2-macroglobulin, also has the ability to bind peptides. The resulting α_2–macroglobulin–peptide complexes (like Gp96–peptide complexes) are able to prime CD8 T-cell responses in vivo (BINDER et al. 2001).

3.3 HSPs Activate Macrophages and Dendritic Cells

To prime naive T cells efficiently at least two signals need to be provided by APCs: the first is the recognition of MHC–peptide complexes by the TCR and its co-receptor. To fulfill this requirement, APCs, and especially DCs, possess unique pathways to shuttle exogenous antigen effectively into the antigen processing pathway leading to the formation of MHC–peptide complexes (RODRIGUEZ et al. 1999). The second signal depends on the interaction of co-stimulatory molecules with, e.g., CD28 expressed by T cells. The absence of co-stimulatory activity is believed to lead to tolerance rather than to priming. The ability to carefully regulate its co-stimulatory molecules renders the DC exceptional among the APCs. Co-stimuli can be provided by several surface markers including CD80, CD86, 4–1BB ligand and others (reviewed in WATTS and DEBENEDETTE 1999) which are up-regulated during DC maturation.

There are numerous sources for the maturation of DCs: bacterial products such as lipopolysaccharide (LPS) or CpG-DNA, pro-inflammatory cytokines such as TNF-α or the engagement of CD40 expressed on APCs by CD40 ligand (CD40L) on T helper cells. It has also been shown that the content of cells undergoing necrosis (but not apoptosis) can activate macrophages and DCs (SAUTER et al. 2000). However, the substances released during necrosis which are responsible for APC stimulation are as yet ill-defined. The question of whether HSPs – one of the most abundant classes of cellular proteins – can induce the activation of macrophages and DCs has been clarified in the past year.

3.4 HSP60 as a Ligand of TLR2 and TLR4

Due to the pronounced immunogenicity of HSP60 (which was noticed well before other functions) immune responses to the HSP60 molecules of invading pathogens are often so prominent that they characterize a given infection (Table 2) (KAUFMANN 1990; LATHIGRA et al. 1991). For example, during infection of mice with *Mycobacterium tuberculosis* about 20% of activated T cells respond to HSP60 of *M. tuberculosis* (KAUFMANN et al. 1987). HSP60 stimulates different

Table 2. Infections with immune response against pathogen's HSP60 (based on Suzue and Young 1996)

Infectious agent	Disease
Bacteria	
Bordetella pertussis	Pertussis
Borrelia burgdorferi	Lyme disease
Brucella abortus	Brucellosis
Chlamydia trachomatis	Blinding trachoma
Coxiella burnetti	Q fever
Helicobacter pylori	Gastritis
Legionella pneumophila	Legionnaires' disease
Mycobacterium leprae	Leprosy
Mycobacterium tuberculosis	Tuberculosis
Treponema pallidum	Syphilis
Fungi	
Aspergillus fumigatus	Aspergillosis
Histoplasma capsulatum	Histoplasmosis

cells, including osteoclasts, monocytes, endothelial cells, keratinocytes and fibroblasts (Kirby et al. 1995; Galdiero et al. 1997; Marcatili et al. 1997; Hinode et al. 1998). The ability to activate innate immune cells such as macrophages and DCs has been documented for chlamydial HSP60 (Kol et al. 1998, 1999). Considerable interest in chlamydial HSP60 and innate immune cell interaction arose as a result of recent progress in atherosclerosis research. Today, atherosclerosis is considered to be a form of chronic inflammation (Ross 1999). In parallel to genetic and epidemiological risk factors, there are hints of a possible infectious origin of coronary heart disease. *Chlamydia pneumoniae* infection in particular has been linked to the pathogenesis of atherosclerosis (Leinonen 2000), although the mechanisms involved are still unclear (Capron 1996). Chlamydial HSP60 has been found to accumulate in atherosclerotic lesions and to stimulate macrophage functions relevant to disease progression, such as the production of proinflammatory cytokines and matrix-degrading metalloproteases (Kol et al. 1998). Interestingly, the human homologue HSP60 has been co-localized with chlamydial HSP60 in vascular lesions and was found to activate endothelial cells, smooth muscle cells, macrophages (Kol et al. 1998, 1999; Chen et al. 1999) and DCs as well (Vabulas et al. 2001). In this way, the so called 'danger' hypothesis received its first molecular substantiation. The danger hypothesis claims that the immune system is oblivious of 'foreign-ness' but rather senses and reacts to danger (Matzinger 1998). Different microbial products such as bacterial LPSs, bacterial lipoproteins, viral double-stranded RNA or bacterial CpG-DNA can act as pathogen derived 'exogenous' danger signals. It was exciting to observe that endogenous HSP60 appears to display characteristics established for 'foreign' substances from pathogens in that HSP60 is able to activate innate immune cells and thus controls the adaptive immune response. During the molecular analysis of the stimulatory activity of endogenous HSP60, significant parallels to microbial stimuli emerged. First became clear that CD14 is essential for HSP60-mediated activation of cells (Kol et al. 2000). CD14 is a high-affinity uptake receptor for LPS (Wright et al. 1990) and for other bacterial products (Pugin et al. 1994). For example, it was shown that the human astrocytoma cell line U373 acquired responsiveness to human HSP60 upon transfection with the CD14. Furthermore, anti-CD14 antibodies blocked HSP60 activation of peripheral blood mononuclear cells (Kol et al. 2000). Originally, the function of CD14 was enigmatic as CD14 is a glycosylphosphatidylinositol-anchored protein without a cytoplasmic part and is thus unable to transmit activation signals. Subsequent cloning of TLR and the clarification of their role in pathogen recognition (elsewhere in this volume) appear to solve this issue. It is now believed that TLR transduce the activation signal and that CD14 acts as a ligand-scavenging co-receptor able to enhance the overall sensitivity of the recognition system (Aderem and Ulevitch 2000).

The involvement of CD14 in HSP60 signaling indicated that TLRs might recognize HSP60. Indeed, murine bone-marrow derived macrophages respond to human HSP60 by the secretion of proinflammatory cytokine TNF-α and by the production of nitric oxide. This response was completely ablated in macrophages from C3H/HeJ mice (Ohashi et al. 2000). C3H/HeJ mice carry a mutation rendering them LPS-unresponsive. Genetic mapping revealed a mutation in the cytoplasmic part of TLR4 (P712H) which explains why the receptor is defective in transducing LPS signals (Poltorak et al. 1998). Although these data implied TLR4 as candidate for the HSP60 signaling receptor, C3H/HeJ mice have mutations in addition. For example, the Ran GTPase is also mutated, and this has unequivocal functional consequences, at least in LPS signaling (Wong et al. 1999; Yuan et al. 2000).

With this caveat in mind HSP60-mediated signaling was analyzed not only in TLR-deficient (knock-out or mutant) mice but also by genetic complementation with TLRs of otherwise HSP60 unresponsive cells (Vabulas et al. 2001). To do so, the HSP60 unresponsive cell line 293T was selected and transfected with different TLR expression constructs. In this system TLR4 conferred reactivity to both bacterial (chlamydial) and human HSP60. Interestingly, only in combination with MD-2 could TLR4 build up a functional receptor as is the case for LPS (Shimazu et al. 1999). In addition, TLR2 was also able to complement HSP60 responsiveness. This was not unexpected because an earlier report showed an impaired HSP60-driven activation of the Chinese hamster ovary cell line known to have a null mutation in TLR2 (Kol et al. 2000).

As discussed above an interesting aspect of HSP biology refers to the receptor-mediated uptake by APC (Arnold-Schild et al. 1999; Castellino et al. 2000; Binder et al. 2000a; Basu et al. 2001). This possibly applies to HSP60 as well, as under serum-free conditions HSP60 molecules become rapidly endocytosed by macrophages (Vabulas et al. 2001). The observed blocking effect of fetal calf serum on the uptake of HSP60 could be explained by component(s) in the serum that scavenge HSP60. It might also be possible that serum withdrawal stresses the cells to such an extent that they alter the expression of membrane proteins necessary for HSP60 internalization. Alternatively, component(s) in serum could compete with HSP60 for the putative receptor on macrophages. The requirement of endocytosis for the stimulatory action of HSP60 was shown using an inhibitor of receptor-mediated endocytosis which specifically affects the downstream activation of JNK (Vabulas et al. 2001).

3.5 Gp96 as a Ligand of TLR2 and TLR4

Features similar to those observed for HSP60 have been described recently for the endoplasmic reticulum-resident chaperone Gp96. This molecule has been demonstrated to induce maturation of human and mouse dendritic cells visualized by the up-regulation of co-stimulatory molecules resulting in the enhanced stimulation of T cells. Furthermore, upon contact with Gp96 DCs release pro-inflammatory cytokines (Singh-Jasuja et al. 2000b). Maturation of DCs by Gp96 is dependent on a signal transduction pathway involving the transcription factor nuclear factor-κB (Basu et al. 2000). Recently, it was shown that TLRs are involved in Gp96-induced DC maturation (Vabulas et al. 2002) in a fashion similar to that reported for HSP60. Bone-marrow derived DCs from TLR4-deficient mice were unresponsive to Gp96. Genetic complementation with TLR4 plus MD2 and also TLR2 conveyed responsiveness to Gp96, but TLR2 knock-out mice did not show an altered response. It is most likely that TLR4 compensates for TLR2 in vivo, whereas Gp96-mediated stimuli rely strictly on the presence of TLR4. As reported for HSP60, blocking receptor-mediated endocytosis by the transglutaminase inhibitor monodansylcadaverine abrogates Gp96-induced DC maturation while LPS activation remains unaffected. One interpretation for this result is that Gp96 might be recruited to endosomes by its endocytosis receptor CD91 where it meets its activation receptors TLR2/TLR4 (Fig. 1). CD91 itself does not possess a typical signaling domain. Once TLRs have been engaged, several MyD88-dependent signaling pathways are switched on, leading to activation of stress-activated protein kinases and Iκ-B kinase (Vabulas et al. 2002). It is intriguing that APCs use the same receptors and signaling cascades to become activated by HSPs as the pathogen-associated molecular patterns (PAMPs) of LPS, bacterial DNA or double-stranded RNA. These results imply that TLRs are not only involved in the innate immune response to exogenous pathogens, but also in the response to HSPs. It is tempting to consider TLRs as danger signal receptors in general. Perhaps TLRs sense occurring cell death by recognizing released HSPs in the extracellular space.

A possible skepticism about the in vivo relevance of HSP-mediated cross-presentation and APC activation could be raised as rather high concentrations of HSP need to be used for DC activation, ranging from 10 to 100 μg/ml in vitro. However, the high immunostimulatory potential of HSPs towards DCs has been demonstrated in mice challenged with HSP. The injection of 1 μg of Gp96 led to the massive infiltration of CD11c-positive DCs into lymph nodes draining the point of injection. These DCs were of the mature phenotype and were highly efficient in the stimulation of T cells (Binder et al. 2000b).

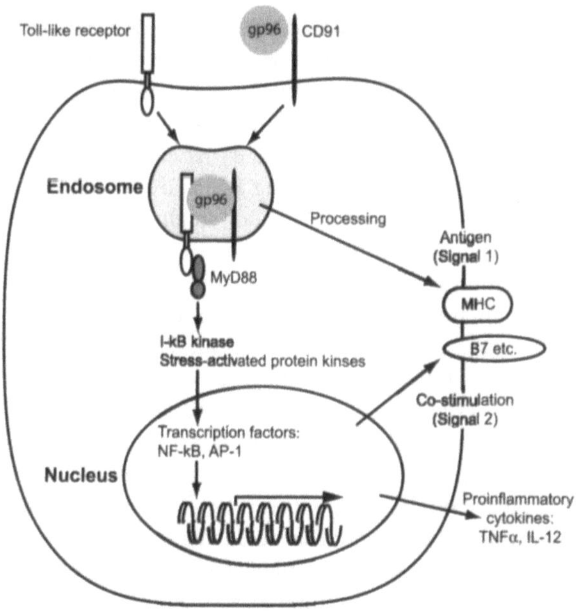

Fig. 1. HSPs interact efficiently with APCs due to the presence of CD91 on the cell surface (in the case of Gp96). Representation of Gp96-associated peptides by MHC molecules (*signal 1*) as well as activation of APCs requires the endocytosis of Gp96 molecules. Activation of APCs occurs via a TLR- and MyD88-dependent pathway and culminates in production of proinflammatory cytokines and up-regulation of costimulatory molecules (*signal 2*)

Taken together, recent reports conclusively show that HSPs alone can act as robust stimulating agents for APCs, especially DCs. The activation of APCs enables DCs to communicate efficiently with T cells resulting in subsequent stimulation of a strong adaptive immune response.

4 HSPs as Necrosis Messengers, HSP Receptors as Sensors for Necrosis

As HSPs are known to be localized inside cells, the source of extracellular Gp96 and other HSPs must be found. Indeed, early experiments identified Gp96 as a cell surface protein on fibroblasts (Pouyssegur and Yamada 1978). However, there is no unequivocal evidence so far for secretion of HSPs. Much more plausible is the argument that HSPs are released as a consequence of necrotic cell death. Indeed, it has been demonstrated that necrotic cells induced by repeated freeze–thaw cycles or by virus infection do release Gp96, HSP90, HSP70 and Calreticulin while apoptotic cells generated by UV radiation do not (Berwin et al. 2001; Basu et al. 2000). Both necrosis and apoptosis, although they do eventually lead to the death of a cell, have very different starting points. Apoptosis, also known as programmed or 'silent' cell death, occurs all the time, especially during the development of an organism and it is a primary 'tool' in the education of the immune system inside the thymus. On the other hand, necrosis is the result of severe stress or trauma, either caused by injury or by infection. Upon necrosis, cellular proteins, including HSPs are released into the surrounding extracellular fluid. APCs in the neighborhood might acquire HSPs and shuttle the associated peptides into the correct processing pathway for presentation on MHC molecules. Simultaneously, HSPs activate APCs, thereby inducing the release of pro-inflammatory cytokines and the up-regulation of co-stimulatory molecules (Singh-Jasuja et al. 2000b; Basu et al. 2000; Asea et al. 2000; Vabulas et al. 2002). The presentation of immunogenic peptides on MHC molecules together with co-stimulatory signals will result in the induction of a T-cell response. The ensuing immune responses would be directed against infected cells in the case of viral infection or would simply elicit a preventive danger response in the case of injury. In the case of apoptosis, however, apoptotic cells or their remnants, the so-called apoptotic bodies, expose phosphatidylserine on the outer leaflets of their membranes. Phosphatidylserine is sensed by the phosphatidylserine receptor present on some macrophages and leads to an anti-inflammatory cytokine response that includes IL-10 and transforming growth factor-β release (Fadok et al. 2000). Although apoptotic cells are endocytosed by APCs to bring about the cross-presentation of these antigens (Albert et al. 1998), co-stimulatory activity is not induced (Sauter et al. 2000). Antigen presentation without a co-stimulus results in tolerance and not in priming. Other receptors such as CD14 or CD36 recognize alternative apoptotic cell-associated ligands and can transmit similar tolerogenic signals to the APC (Devitt et al. 1998). Taken together this model proposes that apoptosis creates a tolerogenic milieu, while necrosis triggers an immunogenic milieu. HSPs may well play a vital role in these processes: HSPs as necrosis messengers, HSP receptors as necrosis sensors.

Experiments identifying CD91 as a receptor for Gp96 revealed that the re-presentation of Gp96-associated peptides after receptor-mediated uptake can be blocked by antibodies against CD91 or by α_2-macroglobulin (Binder et al. 2000a). The latter observation was incorporated into a model for a general role of HSPs in the regulation of APC function. In the blood, uptake of Gp96 (and possibly other HSPs) may be inhibited by α_2-macroglobulin, which is present in the serum in large amounts. In tissues, however, extracellular Gp96 can access CD91 and thus is taken up via receptor-mediated endocytosis leading to the subsequent presentation of its associated peptides on MHC molecules. However, this pathway controls only 'cross-presentation' of shuttled peptides while TLR4 and TLR2 recognition of HSPs controls secretion of pro-inflammatory cytokines and induction of costimuli (signal 2). Interestingly, platelets have been found to possess receptors for Gp96 and to interfere specifically with Gp96-mediated DC activation (whereas DC stimulation by LPS is insensitive to the presence of platelets). Therefore, platelets might be involved in the neutralization of Gp96 and possibly other HSPs in the blood and in wounds (Hilf et al. 2001).

5 Summary

So far TLRs have been described as sensors for 'exogenous' PAMPs believed to be crucial for the initiation of an innate immune response. During evolution these mechanisms developed long before the adaptive immune system evolved. The latest additions to the growing list of ligands for TLRs are HSPs. Interestingly, not only bacterial but also mammalian HSP molecules interact with TLRs (Ohashi et al. 2000; Vabulas et al. 2001; Asea et al. 2000; Vabulas et al. 2002) implying that the exclusive association of TLRs with PAMPs needs to be corrected. Human HSP60 and Gp96 are examples of non-pathogen derived ligands of TLRs. More importantly, Gp96 provides an example of how both the innate and the adaptive immune system can be stimulated by the same molecule released under physiological situations from necrotic cells (Berwin et al. 2001). The importance of these TLR-mediated stimuli for the induction of T helper type 1-dominated immune responses has been observed recently using $MyD88^{-/-}$ mice (Schnare et al. 2001).

Gp96 has kept the (probably HSP60-derived) ability to stimulate APCs non-specifically via TLRs but obtained a new function: to act as a carrier for antigenic peptides and to promote receptor-mediated uptake of antigens by professional APCs (Singh-Jasuja et al. 2000a; Binder et al. 2000a; Basu et al. 2001; Singh-Jasuja et al. 2000b). This unique combination of both features allows the MHC-restricted presentation of antigenic peptides to cells of the adaptive immune system in an immunostimulatory context and enables DCs to act as coordinators of innate and adaptive immune responses. By understanding these mechanisms it will be possible to interfere with the HSP-mediated activation of APCs and to rationally modulate immune responses, either towards immunity or tolerance.

Acknowledgements. H.S. is supported by Grants from the Deutsche Forschungsgemeinschaft (Schi301-2; SFB510 C1) and from the European Union (EC Project QLK3-CT-1999-00064). The authors would like to thank L. Yakes for excellent editorial assistance.

Note added in proof. Recently two papers describing involvement of TLR2/TLR4 in HSP70 signaling appeared: Vabulas et al. (2002) J Biol Chem 277:15107–15112; Asea et al. (2002) J Biol Chem 277:15028–15034.

References

Aderem A, Ulevitch RJ (2000) Toll-like receptors in the induction of the innate immune response. Nature 406:782–787

Agashe VR, Hartl FU (2000) Roles of molecular chaperones in cytoplasmic protein folding. Semin Cell Dev Biol 11:15–25

Albert ML, Pearce SF, Francisco LM, Sauter B, Roy P, Silverstein RL, Bhardwaj N (1998) Immature dendritic cells phagocytose apoptotic cells via alphavbeta5 and CD36, and cross-present antigens to cytotoxic T lymphocytes. J Exp Med 188:1359–1368

Arnold D, Faath S, Rammensee H, Schild H (1995) Cross-priming of minor histocompatibility antigen-specific cytotoxic T cells upon immunization with the heat shock protein gp96. J Exp Med 182:885–889
Arnold-Schild D, Hanau D, Spehner D, Schmid C, Rammensee HG, de la Salle H, Schild H (1999) Cutting edge: receptor-mediated endocytosis of heat shock proteins by professional antigen-presenting cells. J Immunol 162:3757–3760
Asea A, Kraeft SK, Kurt-Jones EA, Stevenson MA, Chen LB, Finberg RW, Koo GC, Calderwood SK (2000) HSP70 stimulates cytokine production through a CD14-dependant pathway, demonstrating its dual role as a chaperone and cytokine. Nature Med 6:435–442
Banchereau J, Briere F, Caux C, Davoust J, Lebecque S, Liu YJ, Pulendran B, Palucka K (2000) Immunobiology of dendritic cells. Annu Rev Immunol 18:767–811
Basu S, Binder RJ, Ramalingam T, Srivastava PK (2001) CD91 is a common receptor for heat shock proteins gp96, hsp90, hsp70, and calreticulin. Immunity 14:303–313
Basu S, Binder RJ, Suto R, Anderson KM, Srivastava PK (2000) Necrotic but not apoptotic cell death releases heat shock proteins, which deliver a partial maturation signal to dendritic cells and activate the NF-kappaB pathway. Int Immunol 12:1539–1546
Berwin B, Reed RC, Nicchitta CV (2001) Virally induced lytic cell death elicts the release of immunogenic grp94/Gp96. J Biol Chem (in press)
Binder RJ, Han DK, Srivastava PK (2000a) CD91: a receptor for heat shock protein gp96. Nat Immunology 1:151–155
Binder RJ, Anderson KM, Basu S, Srivastava PK (2000b) Cutting edge: heat shock protein gp96 induces maturation and migration of CD11c(+) cells In vivo. J Immunol 165:6029–6035
Binder RJ, Karimeddini D, Srivastava PK (2001) Adjuvanticity of alpha(2)-Macroglobulin, an Independent Ligand for the Heat Shock Protein Receptor CD91. J Immunol 166:4968–4972
Bukau B, Horwich AL (1998) The Hsp70 and Hsp60 chaperone machines. Cell 92:351–366
Capron L (1996) Chlamydia in coronary plaques-hidden culprit or harmless hobo? Nature Med 2: 856–857
Castellino F, Boucher PE, Eichelberg K, Mayhew M, Rothman JE, Houghton AN, Germain RN (2000) Receptor-mediated uptake of antigen/heat shock protein complexes results in major histocompatibility complex class I antigen presentation via two distinct processing pathways. J Exp Med 191: 1957–1964
Chen W, Syldath U, Bellmann K, Burkart V, Kolb H (1999) Human 60-kDa heat-shock protein: a danger signal to the innate immune system. J Immunol 162:3212–3219
Devitt A, Moffatt OD, Raykundalia C, Capra JD, Simmons DL, Gregory CD (1998) Human CD14 mediates recognition and phagocytosis of apoptotic cells. Nature 392:505–509
Ellis J (1987) Proteins as molecular chaperones. Nature 328:378–379
Ellis RJ (2001) Macromolecular crowding: an important but neglected aspect of the intracellular environment. Curr Opin Struct Biol 11:114–119
Fadok VA, Bratton DL, Rose DM, Pearson A, Ezekewitz RA, Henson PM (2000) A receptor for phosphatidylserine-specific clearance of apoptotic cells. Nature 405:85–90
Fayet O, Ziegelhoffer T, Georgopoulos C (1989) The groES and groEL heat shock gene products of Escherichia coli are essential for bacterial growth at all temperatures. J Bacteriol 171:1379–1385
Galdiero M, de l'Ero GC, Marcatili A (1997) Cytokine and adhesion molecule expression in human monocytes and endothelial cells stimulated with bacterial heat shock proteins. Infect Immun 65: 699–707
Gao Y, Thomas JO, Chow RL, Lee GH, Cowan NJ (1992) A cytoplasmic chaperonin that catalyzes beta-actin folding. Cell 69:1043–1050
Hendrick JP, Hartl FU (1993) Molecular chaperone functions of heat-shock proteins. Annu Rev Biochem 62:349–384
Hilf N, Singh-Jasuja H, Scherer HU, Schwarzmaier P, Couttefangeas C, Schild H, Rammensee H (2001) A negative regulator for Gp96 induced dendritic cell activation: the ER-resident heat shock protein binds to human platelets. Blood (in press)
Hinode D, Yoshioka M, Tanabe S, Miki O, Masuda K, Nakamura R (1998) The GroEL-like protein from Campylobacter rectus: immunological characterization and interleukin-6 and -8 induction in human gingival fibroblast. FEMS Microbiol Lett 167:1–6
Houry WA, Frishman D, Eckerskorn C, Lottspeich F, Hartl FU (1999) Identification of in vivo substrates of the chaperonin GroEL. Nature 402:147–154
Karlin S, Brocchieri L (2000) Heat shock protein 60 sequence comparisons: duplications, lateral transfer, and mitochondrial evolution. Proc Natl Acad Sci U S A 97:11348–11353
Kaufmann SH (1990) Heat shock proteins and the immune response. Immunol Today 11:129–136

Kaufmann SH, Vath U, Thole JE, Van Embden JD, Emmrich F (1987) Enumeration of T cells reactive with Mycobacterium tuberculosis organisms and specific for the recombinant mycobacterial 64-kDa protein. Eur J Immunol 17:351–357

Kiang JG, Tsokos GC (1998) Heat shock protein 70 kDa: molecular biology, biochemistry, and physiology. Pharmacol Ther 80:183–201

Kirby AC, Meghji S, Nair SP, White P, Reddi K, Nishihara T, Nakashima K, Willis AC, Sim R, Wilson M, et al. (1995) The potent bone-resorbing mediator of Actinobacillus actinomycetemcomitans is homologous to the molecular chaperone GroEL. J Clin Invest 96:1185–1194

Kol A, Bourcier T, Lichtman AH, Libby P (1999) Chlamydial and human heat shock protein 60 s activate human vascular endothelium, smooth muscle cells, and macrophages. J Clin Invest 103: 571–577

Kol A, Lichtman AH, Finberg RW, Libby P, Kurt-Jones EA (2000) Cutting edge: heat shock protein (HSP) 60 activates the innate immune response: CD14 is an essential receptor for HSP60 activation of mononuclear cells. J Immunol 164:13–17

Kol A, Sukhova GK, Lichtman AH, Libby P (1998) Chlamydial heat shock protein 60 localizes in human atheroma and regulates macrophage tumor necrosis factor-alpha and matrix metalloproteinase expression. Circulation 98:300–307

Kubota H, Hynes G, Carne A, Ashworth A, Willison K (1994) Identification of six Tcp-1-related genes encoding divergent subunits of the TCP-1-containing chaperonin. Curr Biol 4:89–99

Lathigra RB, Butcher PD, Garbe TR, Young DB (1991) Heat shock proteins as virulence factors of pathogens. Curr Top Microbiol Immunol 167:125–143

Leinonen M (2000) Chlamydia pneumoniae and other risk factors for atherosclerosis. J Infect Dis 181:S414–S416

Lindquist S (1986) The heat-shock response. Annu Rev Biochem 55:1151–1191

Marcatili A, Cipollaro de l'Ero G, Galdiero M, Folgore A, Petrillo G (1997) TNF-alpha, IL-1 alpha, IL-6 and ICAM-1 expression in human keratinocytes stimulated in vitro with Escherichia coli heat-shock proteins. Microbiology 143:45–53

Matzinger P (1998) An innate sense of danger. Semin Immunol 10:399–415

Mayer MP, Bukau B (1999) Molecular chaperones: the busy life of Hsp90. Curr Biol 9:R322–R325

Ohashi K, Burkart V, Flohe S, Kolb H (2000) Cutting edge: heat shock protein 60 is a putative endogenous ligand of the toll-like receptor-4 complex. J Immunol 164:558–561

Poltorak A, He X, Smirnova I, Liu MY, Huffel CV, Du X, Birdwell D, Alejos E, Silva M, Galanos C, Freudenberg M, Ricciardi-Castagnoli P, Layton B, Beutler B (1998) Defective LPS signaling in C3H/HeJ and C57BL/10ScCr mice: mutations in Tlr4 gene. Science 282:2085–2088

Pouyssegur J, Yamada KM (1978) Isolation and immunological characterization of a glucose-regulated fibroblast cell surface glycoprotein and its nonglycosylated precursor. Cell 13:139–140

Pratt WB (1997) The role of the hsp90-based chaperone system in signal transduction by nuclear receptors and receptors signaling via MAP kinase. Annu Rev Pharmacol Toxicol 37:297–326

Prodromou C, Roe SM, O'Brien R, Ladbury JE, Piper PW, Pearl LH (1997) Identification and structural characterization of the ATP/ADP-binding site in the Hsp90 molecular chaperone. Cell 90:65–75

Pugin J, Heumann ID, Tomasz A, Kravchenko VV, Akamatsu Y, Nishijima M, Glauser MP, Tobias PS, Ulevitch RJ (1994) CD14 is a pattern recognition receptor. Immunity 1:509–516

Rodriguez A, Regnault A, Kleijmeer M, Ricciardi-Castagnoli P, Amigorena S (1999) Selective transport of internalized antigens to the cytosol for MHC class I presentation in dendritic cells. Nature Cell Biol 1:362–368

Ross R (1999) Atherosclerosis–an inflammatory disease. N Engl J Med 340:115–126

Rudiger S, Germeroth L, Schneider-Mergener J, Bukau B (1997) Substrate specificity of the DnaK chaperone determined by screening cellulose-bound peptide libraries. EMBO J 16:1501–1507

Sauter B, Albert ML, Francisco L, Larsson M, Somersan S, Bhardwaj N (2000) Consequences of cell death: exposure to necrotic tumor cells, but not primary tissue cells or apoptotic cells, induces the maturation of immunostimulatory dendritic cells. J Exp Med 191:423–434

Schnare M, Barton GM, Holt AC, Takeda K, Akira S, Medzhitov R (2001) Toll-like receptors control activation of adaptive immune responses. Nature Immunol 2:947–950

Shimazu R, Akashi S, Ogata H, Nagai Y, Fukudome K, Miyake K, Kimoto M (1999) MD-2, a molecule that confers lipopolysaccharide responsiveness on Toll-like receptor 4. J Exp Med 189:1777–1782

Singh-Jasuja H, Toes RE, Spee P, Munz C, Hilf N, Schoenberger SP, Ricciardi-Castagnoli P, Neefjes J, Rammensee HG, Arnold-Schild D, Schild H (2000a) Cross-Presentation of Glycoprotein 96-associated Antigens on Major Histocompatibility Complex Class I Molecules Requires Receptor-mediated Endocytosis. J Exp Med 191:1965–1974

Singh-Jasuja H, Scherer HU, Hilf N, Arnold-Schild D, Rammensee HG, Toes REM, Schild H (2000b) The heat shock protein gp96 induces maturation of dendritic cells and down-regulation of its receptor. Eur J Immunol 30:2211–2215
Srivastava PK, Udono H, Blachere NE, Li Z (1994) Heat shock proteins transfer peptides during antigen processing and CTL priming. Immunogenetics 39:93–98
Strickland DK, Kounnas MZ, Argraves WS (1995) LDL receptor-related protein: a multiligand receptor for lipoprotein and proteinase catabolism. FASEB J 9:890–898
Suto R, Srivastava PK (1995) A mechanism for the specific immunogenicity of heat shock protein-chaperoned peptides. Science 269:1585–1588
Suzue K, Young RA (1996) Heat shock proteins as immunological carriers and vaccines. EXS 77:451–465
Thulasiraman V, Yang CF, Frydman J (1999) In vivo newly translated polypeptides are sequestered in a protected folding environment. EMBO J 18:85–95
Vabulas RM, Ahmad-Nejad P, da Costa C, Miethke T, Kirschning CJ, Hacker H, Wagner H (2001) Endocytosed HSP60 s use toll-like receptor 2 (TLR2) and TLR4 to activate the toll/interleukin-1 receptor signaling pathway in innate immune cells. J Biol Chem 276:31332–31339
Vabulas RM, Braedel S, Hilf N, Singh-Jasuja H, Herter S, Ahmad-Nejad P, Kirschning CJ, da Costa C, Rammensee HG, Wagner H, Schild H (2002) The endoplasmic reticulum-resident heat shock protein Gp96 activates dendritic cells via the toll-like receptor 2/4 pathway. J Biol Chem 277:20847–20853
Wassenberg JJ, Dezfulian C, Nicchitta CV (1999) Receptor mediated and fluid phase pathways for internalization of the ER Hsp90 chaperone GRP94 in murine macrophages. J Cell Sci 112:2167–2175
Watts TH, DeBenedette MA (1999) T cell co-stimulatory molecules other than CD28. Curr Opin Immunol 11:286–293
Wong PM, Kang A, Chen H, Yuan Q, Fan P, Sultzer BM, Kan YW, Chung SW (1999) Lps(d)/Ran of endotoxin-resistant C3H/HeJ mice is defective in mediating lipopolysaccharide endotoxin responses. Proc Natl Acad Sci USA 96:11543–11548
Wright SD, Ramos RA, Tobias PS, Ulevitch RJ, Mathison JC (1990) CD14, a receptor for complexes of lipopolysaccharide (LPS) and LPS binding protein. Science 249:1431–1433
Yaffe MB, Farr GW, Miklos D, Horwich AL, Sternlicht ML, Sternlicht H (1992) TCP1 complex is a molecular chaperone in tubulin biogenesis. Nature 358:245–248
Yuan Q, Zhao F, Chung SW, Fan P, Sultzer BM, Kan YW, Wong PM (2000) Dominant negative down-regulation of endotoxin-induced tumor necrosis factor alpha production by Lps(d)/Ran. Proc Natl Acad Sci USA 97:2852–2857

Subject Index

Current Topics in Microbiology and Immunology

Volumes published since 1989 (and still available)

Vol. 227: **Vogt, Peter K.; Reed, Steven I. (Eds.):** Cyclin Dependent Kinase (CDK) Inhibitors. 1998. 15 figs. XII, 169 pp. ISBN 3-540-63429-0

Vol. 228: **Pawson, Anthony I. (Ed.):** Protein Modules in Signal Transduction. 1998. 42 figs. IX, 368 pp. ISBN 3-540-63396-0

Vol. 229: **Kelsoe, Garnett; Flajnik, Martin (Eds.):** Somatic Diversification of Immune Responses. 1998. 38 figs. IX, 221 pp. ISBN 3-540-63608-0

Vol. 230: **Kärre, Klas; Colonna, Marco (Eds.):** Specificity, Function, and Development of NK Cells. 1998. 22 figs. IX, 248 pp. ISBN 3-540-63941-1

Vol. 231: **Holzmann, Bernhard; Wagner, Hermann (Eds.):** Leukocyte Integrins in the Immune System and Malignant Disease. 1998. 40 figs. XIII, 189 pp. ISBN 3-540-63609-9

Vol. 232: **Whitton, J. Lindsay (Ed.):** Antigen Presentation. 1998. 11 figs. IX, 244 pp. ISBN 3-540-63813-X

Vol. 233/I: **Tyler, Kenneth L.; Oldstone, Michael B. A. (Eds.):** Reoviruses I. 1998. 29 figs. XVIII, 223 pp. ISBN 3-540-63946-2

Vol. 233/II: **Tyler, Kenneth L.; Oldstone, Michael B. A. (Eds.):** Reoviruses II. 1998. 45 figs. XVI, 187 pp. ISBN 3-540-63947-0

Vol. 234: **Frankel, Arthur E. (Ed.):** Clinical Applications of Immunotoxins. 1999. 16 figs. IX, 122 pp. ISBN 3-540-64097-5

Vol. 235: **Klenk, Hans-Dieter (Ed.):** Marburg and Ebola Viruses. 1999. 34 figs. XI, 225 pp. ISBN 3-540-64729-5

Vol. 236: **Kraehenbuhl, Jean-Pierre; Neutra, Marian R. (Eds.):** Defense of Mucosal Surfaces: Pathogenesis, Immunity and Vaccines. 1999. 30 figs. IX, 296 pp. ISBN 3-540-64730-9

Vol. 237: **Claesson-Welsh, Lena (Ed.):** Vascular Growth Factors and Angiogenesis. 1999. 36 figs. X, 189 pp. ISBN 3-540-64731-7

Vol. 238: **Coffman, Robert L.; Romagnani, Sergio (Eds.):** Redirection of Th1 and Th2 Responses. 1999. 6 figs. IX, 148 pp. ISBN 3-540-65048-2

Vol. 239: **Vogt, Peter K.; Jackson, Andrew O. (Eds.):** Satellites and Defective Viral RNAs. 1999. 39 figs. XVI, 179 pp. ISBN 3-540-65049-0

Vol. 240: **Hammond, John; McGarvey, Peter; Yusibov, Vidadi (Eds.):** Plant Biotechnology. 1999. 12 figs. XII, 196 pp. ISBN 3-540-65104-7

Vol. 241: **Westblom, Tore U.; Czinn, Steven J.; Nedrud, John G. (Eds.):** Gastroduodenal Disease and Helicobacter pylori. 1999. 35 figs. XI, 313 pp. ISBN 3-540-65084-9

Vol. 242: **Hagedorn, Curt H.; Rice, Charles M. (Eds.):** The Hepatitis C Viruses. 2000. 47 figs. IX, 379 pp. ISBN 3-540-65358-9

Vol. 243: **Famulok, Michael; Winnacker, Ernst-L.; Wong, Chi-Huey (Eds.):** Combinatorial Chemistry in Biology. 1999. 48 figs. IX, 189 pp. ISBN 3-540-65704-5

Vol. 244: **Daëron, Marc; Vivier, Eric (Eds.):** Immunoreceptor Tyrosine-Based Inhibition Motifs. 1999. 20 figs. VIII, 179 pp. ISBN 3-540-65789-4

Vol. 245/I: **Justement, Louis B.; Siminovitch, Katherine A. (Eds.):** Signal Transduction and the Coordination of B Lymphocyte Development and Function I. 2000. 22 figs. XVI, 274 pp. ISBN 3-540-66002-X

Vol. 245/II: **Justement, Louis B.; Siminovitch, Katherine A. (Eds.):** Signal Transduction on the Coordination of B Lymphocyte Development and Function II. 2000. 13 figs. XV, 172 pp. ISBN 3-540-66003-8

Vol. 246: **Melchers, Fritz; Potter, Michael (Eds.):** Mechanisms of B Cell Neoplasia 1998. 1999. 111 figs. XXIX, 415 pp. ISBN 3-540-65759-2

Vol. 247: **Wagner, Hermann (Ed.):** Immunobiology of Bacterial CpG-DNA. 2000. 34 figs. IX, 246 pp. ISBN 3-540-66400-9

Vol. 248: **du Pasquier, Louis; Litman, Gary W. (Eds.):** Origin and Evolution of the Vertebrate Immune System. 2000. 81 figs. IX, 324 pp. ISBN 3-540-66414-9

Vol. 249: **Jones, Peter A.; Vogt, Peter K. (Eds.):** DNA Methylation and Cancer. 2000. 16 figs. IX, 169 pp. ISBN 3-540-66608-7

Vol. 250: **Aktories, Klaus; Wilkins, Tracy, D. (Eds.):** Clostridium difficile. 2000. 20 figs. IX, 143 pp. ISBN 3-540-67291-5

Vol. 251: **Melchers, Fritz (Ed.):** Lymphoid Organogenesis. 2000. 62 figs. XII, 215 pp. ISBN 3-540-67569-8

Vol. 252: **Potter, Michael; Melchers, Fritz (Eds.):** B1 Lymphocytes in B Cell Neoplasia. 2000. XIII, 326 pp. ISBN 3-540-67567-1

Vol. 253: **Gosztonyi, Georg (Ed.):** The Mechanisms of Neuronal Damage in Virus Infections of the Nervous System. 2001. approx. XVI, 270 pp. ISBN 3-540-67617-1

Vol. 254: **Privalsky, Martin L. (Ed.):** Transcriptional Corepressors. 2001. 25 figs. XIV, 190 pp. ISBN 3-540-67569-8

Vol. 255: **Hirai, Kanji (Ed.):** Marek's Disease. 2001. 22 figs. XII, 294 pp. ISBN 3-540-67798-4

Vol. 256: **Schmaljohn, Connie S.; Nichol, Stuart T. (Eds.):** Hantaviruses . 2001, 24 figs. XI, 196 pp. ISBN 3-540-41045-7

Vol. 257: **van der Goot, Gisou (Ed.):** Pore-Forming Toxins, 2001. 19 figs. IX, 166 pp. ISBN 3-540-41386-3

Vol. 258: **Takada, Kenzo (Ed.):** Epstein-Barr Virus and Human Cancer. 2001. 38 figs. IX, 233 pp. ISBN 3-540-41506-8

Vol. 259: **Hauber, Joachim, Vogt, Peter K. (Eds.):** Nuclear Export of Viral RNAs. 2001. 19 figs. IX, 131 pp. ISBN 3-540-41278-6

Vol. 260: **Burton, Didier R. (Ed.):** Antibodies in Viral Infection. 2001. 51 figs. IX, 309 pp. ISBN 3-540-41611-0

Vol. 261: **Trono, Didier (Ed.):** Lentiviral Vectors. 2002. 32 figs. X, 258 pp. ISBN 3-540-42190-4

Vol. 262: **Oldstone, Michael B.A. (Ed.):** Arenaviruses I. 2002, 30 figs. XVIII, 197 pp. ISBN 3-540-42244-7

Vol. 263: **Oldstone, Michael B. A. (Ed.):** Arenaviruses II. 2002, 49 figs. XVIII, 268 pp. ISBN 3-540-42705-8

Vol. 264/I: **Hacker, Jörg; Kaper, James B. (Eds.):** Pathogenicity Islands and the Evolution of Microbes. 2002. 34 figs. XVIII, 232 pp. ISBN 3-540-42681-7

Vol. 264/II: **Hacker, Jörg; Kaper, James B. (Eds.):** Pathogenicity Islands and the Evolution of Microbes. 2002. 24 figs. XVIII, 228 pp. ISBN 3-540-42682-5

Vol. 265: **Dietzschold, Bernhard; Richt, Jürgen A. (Eds.):** Protective and Pathological Immune Responses in the CNS. 2002. 21 figs. X, 278 pp. ISBN 3-540-42668-X

Vol. 266: **Cooper, Koproski (Eds.):** The Interface Between Innate and Acquired Immunity, 2002, 15 figs. XIV, 116 pp. ISBN 3-540-42894-1

Vol. 267: **Mackenzie, John S.; Barrett, Alan D. T.; Deubel, Vincent (Eds.):** Japanese Encephalitis and West Nile Viruses. 2002. 66 figs. X, 418 pp. ISBN 3-540-42783-X

Vol. 268: **Zwickl, Peter; Baumeister, Wolfgang (Eds.):** The Proteasome-Ubiquitin Protein Degradation Pathway. 2002, X, 213 pp. ISBN 3-540-43096-2

Vol. 269: **Koszinowski, Ulrich H.; Hengel, H. (Eds.):** Viral Proteins Counteracting Host Defenses. 2002, XII, 328 pp. ISBN 3-540-43261-2

MIX
Papier aus verantwortungsvollen Quellen
Paper from responsible sources
FSC® C105338

FSC
www.fsc.org

If you have any concerns about our products,
you can contact us on
ProductSafety@springernature.com

In case Publisher is established outside the EU,
the EU authorized representative is:
Springer Nature Customer Service Center GmbH
Europaplatz 3, 69115 Heidelberg, Germany

Printed by Libri Plureos GmbH
in Hamburg, Germany